NONLINEAR MODELS FOR ARCHAEOLOGY AND ANTHROPOLOGY

In memory of Ilya Prigogine
1917-2003

Nonlinear Models for Archaeology and Anthropology

Continuing the Revolution

Edited by

CHRISTOPHER S. BEEKMAN
University of Colorado at Denver and Health Sciences Center, USA

WILLIAM W. BADEN
Indiana University – Purdue University Fort Wayne, USA

ASHGATE

Published by
Ashgate Publishing Limited
Gower House
Croft Road
Aldershot
Hampshire GU11 3HR
England

Ashgate Publishing Company
Suite 420
101 Cherry Street
Burlington, VT 05401-4405
USA

Ashgate website: http://www.ashgate.com

British Library Cataloguing in Publication Data
Nonlinear models for archaeology and anthropology :
continuing the revolution
1. Social archaeology 2. Nonlinear systems
I. Beekman, Christopher Stockard II. Baden, William W.
930.1'01

Library of Congress Cataloging-in-Publication Data
Nonlinear models for archaeology and anthropology : continuing the revolution / edited by Christopher S. Beekman and William W. Baden.
p. cm.
Includes bibliographical references and index.
ISBN 0-7546-4319-0
1. Archaeology--Mathematical models. 2. Nonlinear systems. I. Beekman, Christopher Stockard. II. Baden, William W.

CC80.6.N66 2005
930.101'51--dc22

2005024902

ISBN 0 7546 4319 0

Printed and bound by Athenaeum Press Ltd,
Gateshead, Tyne & Wear.

Contents

List of Figures

List of Tables

List of Contributors

Robert L. Axtell, Center on Social and Economic Dynamics, The Brookings Institution, raxtell@brookings.edu

I am co-author with Joshua Epstein of *Growing Artificial Societies: Social Science from the Bottom Up* (MIT Press, 1996) and creator of the Sugarscape software on which the book is based. I have been affiliated with the Santa Fe Institute since a graduate student and am currently an External Faculty Member. Together with Brookings colleagues and archaeologists George Gumerman and Jeff Dean, I have worked on an agent based model of the Anasazi in the Long House Valley area of Black Mesa, work published in the *Proceedings of the National Academy of Sciences* and reviewed in *Nature*. I am co-founder of the Center on Social and Economic Dynamics at Brookings, a research center dedicated to development of agent based modelling techniques in the social sciences. Most recently I have been applying agent techniques to problems in economics, including the self-organization of markets and the emergence of firms, and recently finished working on a book entitled *Artificial Economies of Adaptive Agents: The Multi-Agent Systems Approach to Economics*.

William W. Baden, Institutional Research and Analysis, Indiana University – Purdue University Fort Wayne, baden@ipfw.edu

Starting academic life as a physics student with its strong computational emphasis ensured my trajectory once I converted to Anthropology. While a graduate student in the Pre-Google era I literally stumbled upon Nicolis and Prigogine's 1977 self-organization volume on a library shelf. A visiting lecture by Richard N. Adams further inspired me to incorporate these concepts in my dissertation research. The topic was derived from discussions between fellow graduate students about the practical aspects of growing maize. We had all grown up on farms and agreed that agriculture was not inherently self-sustaining. Proving that to 'the city folk' would become the research challenge. Years later this work was resurrected when I crossed paths with Chris Beekman. Our shared interest in the unachieved potential of the Complexity Sciences led us to organize this text. As for the maize, I continue to dedicate my research to all of us that 'know how to hoe a row'.

Christopher S. Beekman, Department of Anthropology, University of Colorado at Denver and Health Sciences Center, christopher.beekman@cudenver.edu

After a graduate education (Ph.D. Vanderbilt University 1996) focused heavily on political economic and humanistic approaches in ancient Mesoamerica, William Baden sparked my interest in complexity while we were both visiting faculty at Indiana University – Purdue University Fort Wayne. I have since made a point of orienting my field projects in Jalisco, Mexico, towards addressing current issues of agency and political behaviour as well as laying the groundwork for a planned computer simulation in the region that more explicitly combines ecological and social factors.

Carole L. Crumley, Department of Anthropology, University of North Carolina, crumley@unc.edu

I have long had an interest in both science and the humanities, with consecutive undergraduate majors in classics, geology, and anthropology (B.A. University of Michigan, Ann Arbor 1966). Doctoral work in North American archaeology (M.A. University of Calgary 1967) and in paleoclimatology, ecology, and Iron Age (Celtic) archaeology (Ph.D. University of Wisconsin, Madison 1972) turned me toward the long-term study of a region in Burgundy, France. I soon found that the current theoretical paradigm in archaeology does not serve the range of forms that power can take in human societies, and first published on the idea of heterarchy in 1979. Heterarchy's origins in brain research and artificial intelligence, coupled with my interest in Earth systems, led me to contemporary complex systems research. I teach state formation, historical ecology, and ethnohistory at the University of North Carolina, Chapel Hill.

Robert J. Hommon, Pacific Area Archaeologist, National Park Service, rob_hommon@nps.gov

I am Senior Cultural Resource Scientist and Archaeologist for the Pacific West Region, Honolulu Office, National Park Service. My interest in the application of complexity theory to human culture began to emerge when reading James Gleick's popular book, *Chaos*, while preparing a paper on Hawaiian archaeology in 1992. I am currently working on a book that will expand on ideas presented in my contribution to this collection.

J. Stephen Lansing, Department of Anthropology, University of Arizona, Slansing7@aol.com

I am professor of anthropology at the University of Arizona with an adjunct appointment in ecology and evolutionary biology, and a research professor at the Santa Fe Institute. I became interested in complexity theory when I gave a talk at the Santa Fe Institute about Balinese water temples, and was asked whether water temple networks could self-organize.

Tammy Stone, Department of Anthropology and Associate Dean, College of Liberal Arts and Sciences, University of Colorado at Denver and Health Sciences Center, tammy.stone@cudenver.edu

I received my Ph.D. from Arizona State University in 1992 and work primarily in the puebloan Southwest with issues of identity, factionalism, and political alliance formation. My interest in nonlinear modelling grew out of my interest in information processing and information as a source of power. As such, I was influenced by the work of Sander van der Leeuw on information flows.

T.J. Wilkinson, Department of Archaeology, University of Edinburgh, Tony.Wilkinson@ed.ac.uk

Following graduate work in high arctic geomorphology I discovered archaeology. Over the past 30 years, as a result of this encounter I have been conducting landscape surveys in the Middle East and the UK. The resultant projects have been aimed at retrieving archaeological data at a regional scale in order to reconstruct and analyze past demographic trends, settlement change, and local agricultural economies. By combining archaeological and landscape surveys with ancient historical records I have been able to document and examine interactions between human populations and environmental change over extended periods of time as well as to establish some of the basic building blocks of the early state. A long term interest in landscape archaeology led to various attempts to re-construct settlement land use systems in the Near East, and it was the inadequacies of earlier models that contributed to my interest in agent based models. Such interests were made more concrete when I was at the Oriental Institute, University of Chicago. There I was introduced to members of a modelling group based within the DIS division of Argonne National Laboratory, Illinois. This led to a productive collaboration between myself and other colleagues at the Oriental Institute with John Christiansen at the ANL. Currently I am a co-principle investigator (with McGuire Gibson and John Christiansen) of a programme that is developing systems of dynamic modelling in order to simulate Bronze Age land use and settlement development in the Near East in the context of environmental fluctuations. Funded by a recent grant from the National Science Foundation (USA), and based around techniques of object based modelling developed at Argonne National Laboratory, Illinois, the MASS Project enables us to simulate human behaviour and crop production under a wide variety of environmental, social and economic conditions.

Preface

This volume owes its genesis to a session on 'The Application of Nonlinear Systems Theory to Archaeology', originally held at the annual meeting of the American Anthropological Association in New Orleans in 2002. The speakers at the session were unanimously interested in seeing publication of articles expanding the conference topics, and the editors duly went to work to secure a publisher. Tony Wilkinson graciously agreed to provide an additional chapter to round out the collection. Steve Lansing, who unfortunately could not participate in the session, enthusiastically agreed to provide closing discussion for this volume. Other conference participants who could not be part of this final volume are Tom Abel, R. Alexander Bentley, Herbert Maschner, and Suzanne Spencer-Wood, but some of their nonlinear perspectives have since seen the light of day elsewhere.

Christopher S. Beekman
Denver

William W. Baden
Fort Wayne

Acknowledgements

The editors have benefited from discussions with Timothy Kohler, Mark Lake, Henry Wright and members of the complex systems group at the University of Michigan. We would like to thank Alison Kirk, Carolyn Court, Pam Bertram, Donna Hamer, Gemma Lowle, and Emily Poulton of Ashgate Publishing for their help in bringing this volume to completion. Pam Zepp was instrumental in helping us achieve mastery of word processing intricacies. Figure 3.1 was used with permission from Elsevier Ltd. Finally we would like to thank our wives, Kathy and Marla, for their patience throughout the process.

Chapter 1

Continuing the Revolution

Christopher S. Beekman
University of Colorado at Denver and Health Sciences Center

William W. Baden
Indiana University – Purdue University Fort Wayne

Introduction

This volume presents theoretical perspectives and case studies extending nonlinear systems theory (commonly called complexity, chaos theory, or dynamical systems) to archaeological studies of politics and economy. Nonlinear systems theory has taken the scientific world by storm in the past two decades with promises to sweep away reductionistic approaches while retaining a scientific and experimental perspective. This revolution has taken place even as the social sciences have struggled to rebuild after the devastating critiques of Post-Modernism. The two trends collided in the 1990s, resulting in a burst of publications in sociology, political science, and anthropology that explored the possibilities that the new science offered. Archaeological applications have developed more slowly, and have not explored as widely. We assembled the contributors for this volume because it appeared to us that nonlinear systems theory was being interpreted and applied in a very limited way by archaeologists. In the course of discussions with colleagues, we found different concepts of how analysis should proceed and significantly different characterizations of what nonlinear dynamics might mean for archaeology.

Is it a 'Post-Modern Science' attacking science from without? Is it a revolution from within that redefines what science itself means? Does complexity constitute a consistent theoretical package, or is it better understood as a cluster of methods capable of being used by different theoretical schools? Nonlinear research encompasses a more polyglot range of concepts than is usually appreciated, and it is very unclear whether we can truly make pronouncements beginning with 'chaos theory states…'. Continued exploration of these concepts is necessary. This volume presents contributions by archaeologists from both sides of the Atlantic and with different theoretical backgrounds seeking to continue the revolution and extend nonlinear systems to new areas of archaeological inquiry, such as social and political organization, and subsistence and political economy.

Nonlinear Systems: A Brief History

Unlike many theoretical approaches used in archaeology, nonlinear systems theory was not so much borrowed from other disciplines as it emerged more or less independently from research in both the physical and social sciences. Research in physics, chemistry, economics, meteorology, biology, and other fields led to a recognition that many types of systems could not be studied using the intentionally reductionistic approach that has been the guiding principle of science since Sir Isaac Newton. In order to better understand energy, motion, disease, and other familiar topics, traditional science has proceeded by breaking the problem down into increasingly smaller parts that could be studied in isolation from the whole. This allowed experimental manipulation of different variables so as to isolate the effects of each one.

The successes of this approach have been numerous and need not be repeated here. But as researchers have tackled more and more difficult problems, the reductionistic model has become less useful. There has been a growing realization that studies of closed systems are often unrealistic, and that assumptions of system equilibrium must be left behind if we are to study more complex systems where the interaction of many components leads to quite unexpected emergent properties. Science as a whole has, in a sense, caught up with the kinds of systems that social scientists have had to deal with for years – open, fluctuating, unpredictable, yet patterned – and has begun to develop conceptual tools for understanding them.

The research that comes under the heading of nonlinear systems is diverse. Related research findings in meteorology, mathematics, and geometry (e.g., Lorenz 1963; Mandelbrot 1977; May 1976) were baptized with the term Chaos theory in the 1960s, encompassing deterministic systems (i.e., based on equations that behave predictably over short periods of time) that nonetheless eventually become unpredictable due to the equations' sensitive dependence upon initial conditions. In other words the relationships between parts are understood, but prediction fails at the level of the entire system due to the complexity of all the interrelationships and our inability to observe the initial conditions with sufficient precision. The canonical example is Lorenz's experiment with weather prediction. Lorenz discovered that an extraordinarily small, seemingly insignificant, variation in the initial input of one parameter (from 0.506127 to 0.506) into his computer program resulted in an eventual divergence between two predictions that were expected to be exactly identical (Gleick 1987:16). This variation related to the arbitrary precision of his model's input, i.e., what he could measure coupled with the accuracy of his observations and the computer's ability to maintain that precision. From a statistical perspective these precision issues produce what appear to be random fluctuations in the model's trajectory. The mechanisms for the weather are highly quantifiable and relatively well-known, but their sensitivity to our ability to accurately measure or observe them means that predictions can only be made a short distance into the future (up to the prediction horizon), after which the potential variations become too great. Despite the impossibility of ever adequately measuring initial conditions (Heisenberg 1927; Nicolis 1995:58-61) or making medium or long term predictions, recurring trends around topological attractors allow retrodictive analysis and explanation. We can typically explain the weather in hindsight, even if we are unable to effectively predict it. Another significant element of classical chaos theory is the concept of bifurcations. Analysis may well identify points in time when change is almost certain to occur, but it is often impossible to determine which of the possible new trajectories will predominate due to apparent elements of randomness. Historical pathways and contingency are therefore deeply rooted elements of chaos theory, despite the presence of the often misinterpreted term 'deterministic' which reminds us that even the most irregular trajectory is defined by causal forces.

Another related field of inquiry that developed in mathematics is Catastrophe theory (Poston and Stewart 1979; Thom 1975), largely a creation of the 1970s. This is closely related to chaos, and largely focuses on how even systems with very few variables can show surprising transformations and nonlinear trajectories depending upon their interaction. Its abstract mathematical and topological nature tended to prohibit its use in real world situations, where it became extremely difficult to quantify any system in terms of just a few variables.

Complexity or Complex systems theory (Pines 1987; Cowan, et al. 1994) emerged in the 1980s and shares many elements with the previous approaches, but has de-emphasized deterministic chaos and instead focuses on how the mutual interaction of many agents or variables can lead to unexpected emergent properties within the broader system. Complex systems theorists therefore explicitly reject the intentionally reductionistic perspective associated with the Newtonian approach to science and have instead tried to develop methods to study entire systems. A major element here has therefore been the imprecise but nonetheless powerful concept of emergence, colloquially described as 'the whole is more than the sum of its parts'. That is, systems cannot be described through the aggregation of their components, but require an entirely new description at the level of the system.

In a terminological shift, complexity researchers have appropriated the term Chaos to contrast unpatterned systems (not strictly true in the earlier definitions of chaos theory) as an opposing pole to Order – a situation of low potential for change (Langton 1990). *Complex* systems (or Anti-Chaos in the most recent parlance – Shermer 1995; Lansing 2003) lie in the boundary zone between these extremes, within which there are forms of temporary stability that emerge through the interaction of internal components. The complex systems approach has been closely linked to the extensive use of computer simulation due to the amount of data and number of variables being analyzed, and the field is significantly related to issues of adaptive learning and artificial intelligence. The best known of these simulations are the agent based models or artificial societies, in which simulators have attempted to model phenomena 'from the bottom up', concentrating on simple rules that nonetheless produce complex behaviour and emergent properties at the level of the group (Langton 1988, 1994; Langton, et al. 1992; Epstein and Axtell 1996).

Much of the impetus for the coalescence of research into complex systems theory came from the influential research of Ilya Prigogine into dissipative structures and self-organization over the prior few decades (e.g., Glansdorff and Prigogine 1971; Nicolis and Prigogine 1977, 1989; Prigogine, et al. 1972a, 1972b). Briefly, his research recognized that pivotal physical laws such as the second law of thermodynamics (the pessimistic law referring to the unavoidable loss of energy) only worked in a closed system, and failed to accurately describe behaviour in an open system in which energy could be introduced from the outside. The introduction of energy produced perturbations to the system and caused it to self-organize into complex patterns far from equilibrium. Common examples of these dissipative systems from the physical world include fluid dynamics and other phase transitions. But the perceptive Prigogine quickly saw the application to biological evolution (particularly the emergence of animate forms from inert matter) and even social change (popularized in Prigogine and Stengers 1984).

Prigogine's own inconsistent use of the terms Chaos, Complexity, and Stability theory to refer to his work reflects the rather loose approach to this material that most practitioners have taken. For this reason, in this volume we prefer the one aspect that all these approaches hold in common – their focus on nonlinear systems. Less catchy perhaps, but more accurate and inclusive. By nonlinear systems we mean those systemic environments in which the number of variables or the relationships between them are so structured as to lead to quite unpredicted and emergent properties. Analysis tends to be oriented towards retrodictive explanation, not prediction, and the simplistic theoretical dichotomies that archaeologists are so attracted to, such as materialism vs. idealism, functionalism vs. conflict, agency vs. structure, or intention vs. adaptation, become difficult to maintain. As opposed to the systems theory of decades past, nonlinear systems are open and subject to varying degrees of stability. Rather like the much touted landscape archaeology of recent years, nonlinear systems invite parallel analysis at multiple scales (see McGlade 2003).

Many prominent social scientists have weighed in on the importance of nonlinear studies for human societies (e.g., Adams 1988; Luhmann 1986, 1995; Maturana and Varela 1980; Wallerstein 1997). Textbooks specifically for use in the social sciences have appeared (Brown 1995; Byrne 1998; Guastello 1995; Marion 1999), numerous collections of applications have been assembled (Albert 1995; Arthur and Arrow 1994; Bertuglia, et al. 1998; Bütz 1997; Ellis and Newton 2000; Eve, et al. 1997; Kiel and Elliott 1996; Leydesdorff and van den Besselaar 1994; Milanovic 1997, 2002; Prietula, et al. 1998; Richards 2000; Schieve and Allen 1982), and a listing of individual studies would probably fill this introduction, although we might single out a few (Allen 1997a; Artigiani 1989; Gilbert 1995; Harvey and Reed 1994, 1996; Lansing 1991; Lansing and Kremer 1993; Leydesdorff 2001; Reed and Harvey 1992; Shermer 1995) as fruitfully exploring the links between existing social theory and nonlinear approaches. But the dynamism in nonlinear systems theory may perhaps have most to offer archaeology, as the social science most deeply involved in the issue of social change over time.

Archaeology and Nonlinear Systems

The earliest interest in nonlinear modelling for archaeology was primarily amongst those European and American researchers who had drifted away from the ecological functionalism, or Processualism, of the 1960s and 1970s to explore the utility of General Systems Theory (Maruyama 1963; von Bertalanffy 1951; Wiener 1948). Concepts broader than adaptation were considered, such as the role of information in social change, and archaeologists such as David Clarke, Colin Renfrew, Sander van der Leeuw, and Ezra Zubrow were of central importance, both in the work they did and the work they inspired in their students (e.g., Clarke 1972, 1978; Renfrew and Cooke 1979; Van der Leeuw 1981a; Van der Leeuw and McGlade 1997a; Van der Leeuw and Torrence 1989; Zubrow 1985). A genuine problem remained in that the rare applications of nonlinear approaches at this time were very simplistic and abstract (Renfrew's application of catastrophe theory to the Maya collapse used 3 variables), and it remained quite difficult to determine exactly how to measure or even identify the important variables in a human system. Hence variants such as catastrophe theory enjoyed an extremely brief period of popularity (e.g., Renfrew 1978, 1979).

Ilya Prigogine was a distinct source from which the interest in nonlinear systems spread to archaeology. Surely encouraged by his own wide range of interests (see Prigogine 1998 to see what we mean), colleagues of his began to apply dissipative structures to the social sciences (e.g., Schieve and Allen 1982; Adams 1988). Eventually, the intellectual descendents of this line carried the approach further to archaeology (e.g., McGlade 1990, 1999). Besides his work out of Brussels, Prigogine also spent time at the University of Texas at Austin, and Americans became introduced to his work more directly (perhaps first mentioned by Blanton and Kowalewski 1982:15). This southern focus, European in origin, independently influenced the early work of some of the contributors to this volume (Baden 1987, 1995; Stone 1999).

Nonlinear systems approaches had an unclear relationship to other theoretical schools in archaeology. It was the eclectic Jonathan Friedman (1982) who would make reference to Prigogine's work in the context of punctuated equilibrium and other rapid transformational models from the physical and biological sciences as part of a larger argument – that Processual archaeologists, in their desire to put their discipline on a scientific footing, had chosen idealized and discredited equilibrium-based theories as their model of science. Nonlinear systems research thus went from being an approach parallel to Processualism to an independent source of critique.

But interest in these approaches may have failed to catch on more widely because most of these applications focused on the analysis of entire systems (see particularly the contributions in Van der Leeuw and McGlade 1997a), even as archaeology as a discipline was turning away from macro-scale analysis. Regardless of whether it caught on immediately or not, Prigogine's work spawned some of the earliest direct archaeological applications of nonlinear systems since Renfrew and Zubrow (Baden 1987; McGlade 1990). It is also members of this loosely defined 'Prigoginian' line that will make attempts to confront the Post-Processual critique, by addressing issues of power and top-down political control (McGlade 1997).

Significant changes occurred in the field in the late 1980s and 1990s, due to two major related trends. The first was the rapid growth in computing power, allowing the development of considerably more complex simulations, particularly agent based modelling. The second was the development of institutional infrastructure in the form of the Santa Fe Institute (SFI) and other research organizations within or parallel to university settings. We suspect that agent based modelling created a fascination for archaeologists that earlier system-wide analyses of complexity could not, and it brought the scale of modelling down to the level of individuals just as other theoretical approaches in archaeology were doing the same. But we must also consider the training in technical methods that the new infrastructure made possible. Few archaeologists leave graduate school with a mathematical background beyond that required to carry out multivariate statistical analyses, and interdisciplinary institutions such as SFI provide an entirely new training opportunity. The simulations of Anasazi culture change by Timothy Kohler, Jeffrey Dean, George Gumerman, and their colleagues are the best known examples of archaeological collaborations to come out of SFI (Dean, et al. 2000; Kohler and Van West 1996; Kohler, et al. 2000). But this highly prominent organization (positively treated in Waldrop 1992 and less so in Helmreich 1998) served as host to a series of other archaeologists over the 1990s and 2000s such as Robert Hommon, Mark Lehner, Suzanne Spencer-Wood, Tony Wilkinson, and Henry Wright (see also the contributions in Kohler and Gumerman 2000). Some have worked on computer simulation while others have chosen to apply complex systems more metaphorically, and examples of their work appear in this volume.

Certain prominent themes run through much of the research at SFI whether practiced by archaeologists or by representatives of other disciplines. Human societies, flocks of birds, the economy, sand piles, the brain, immune systems, ecologies, ant colonies, and pre-organic molecular formations are all conceptualized as Complex Adaptive Systems (CAS) sharing many features in common (Lansing 2003). That is, much of the self-organization of these systems is described as occurring under adaptive pressures towards greater fitness. Modelling has often, though not exclusively, made use of computational mechanisms such as modelling strings of cultural attributes through genetic algorithms (Holland 1998), adaptive interaction through game theoretic exchanges (Axelrod 1984), the creation of fitness landscapes to represent optimal and sub-optimal solutions (Kauffman 1999), artificial life studies involving adaptive learning (Langton 1988, 1994; Langton, et al. 1992), and other tools with strong ties to some form of Darwinian selection. Even when religious concepts or symbols are given an equally important role in modelling behaviour or communication, they are often operationalized again through programming methods that follow a selectionist structure.

Thus CAS is a distinctive approach that does not represent the entire gamut of theories currently practiced in archaeology. That in turn suggests that there is only room here for certain theoretical approaches to society. SFI has been very inclusive in its efforts to cross traditional disciplinary boundaries. But research concentrating on the similarities between complex systems may fail to appreciate the differences between them, which can be just as significant for our understanding. Adaptivist theory is perfectly legitimate as a theoretical approach, but the rather exclusive focus on the concept suggests that perhaps other theoretical schools are not compatible with CAS. This is an implication to be considered as we look at other areas.

Research on human societies and nonlinear systems in Europe traces distinct trajectories. Themes that run through social simulation work in western Europe by researchers like Nigel Gilbert and Jim Doran (Gilbert and Conte 1995; Gilbert and Troitzsch 1999; Gilbert and Doran 1994) include religion, belief systems, irrationality, and social action. Social theorists such as Anthony Giddens and Fernand Braudel are cited alongside Ilya Prigogine. This greater interest in the social leads to applications that look quite different from those recently appearing in the United States (e.g., Doran 2000; Doran, et al. 1994; various in Van der Leeuw and McGlade 1997a) and greater weight to the notion that social systems are qualitatively different from other complex systems. Archaeologists working within this intellectual milieu have often chosen to use nonlinear concepts as metaphors to frame a verbal analysis rather than develop computer simulations (e.g., Bintliff 1997, 1999a, 2003).

The last five years have seen the range of practical applications of nonlinear systems theory to archaeology continue to expand. Fractal analysis has been applied to Maya and Central Plains settlement patterns (Blakeslee 2002; Brown and Witschey 2003; Brown, et al. 2005) and urbanism at Teotihuacan (Oleschko, et al. 2000), technological innovations have been examined through the metaphorical application of dynamical systems and emergence (Roux 2003), self-organized criticality has been used to analyze pottery style (Bentley and Maschner 2001), and nonlinearity has been applied to human organization (Crumley 2001). Nonlinear systems theory has been discussed (R. McC. Adams 2001) and applied (Yoffee 2005) in more metaphorical terms to Mesopotamian social evolution, and theoretical links between Feminist theory and complexity have been outlined (Spencer-Wood 2000). R. Alexander Bentley and Herbert Maschner have published a diverse collection of articles (Bentley and Maschner 2003) examining complex systems in archaeology and history. Their contributors examined settlement patterns, chronology, and social networks, and came much closer than previous volumes to what we wanted to assemble here.

Despite (or because of) its growing profile, nonlinear systems theory has had its detractors. Helmreich (1998) examined the Santa Fe Institute from a deconstructionist perspective, pointing out that white male heterosexuals make up the majority of complex systems researchers and arguing that this has led to a narrower view of social modelling than might otherwise be the case (e.g., use of the metaphor of biological reproduction). Khalil (1995) argues that the work of Prigogine and others is inappropriate for human societies, where intention distinguishes human organization from quasi-cyclical phenomena such as those modelled from the natural world. Offhand comments by prominent archaeologists such as Adam Smith (2003:104) and private communications with colleagues have alluded to what many perceive as the lack of complexity in 'complex' systems research. Much of this comes from theoretical perspectives other than the selectionist approaches welcomed within prominent centres like SFI. But recently there have also been critiques from within. McGlade (2003) has expressed concern that nonlinear dynamics embraces a wider array of concepts, methods, and implications than appreciated by some practitioners. Among other things, he argues for the construction of multiple models of the same system from different perspectives, for more attention to differences across scales, and for a greater appreciation of the fact that we are constructing *tools*, not representations of reality. Mark Lake (2005) has recently argued that the utility of computer simulation of complex systems may be more limited than once thought. It may be most useful for sensitivity analysis and specific methodological questions, but it has produced far less in the area of theory. These critiques, especially from those in the thick of current research, point to the need for a more detailed examination of nonlinear systems' applicability to the archaeological record, and a careful appraisal of current approaches.

This Volume and the Contributors

A brief exposition of nonlinear studies in Thomas Kuhn's (1970) terms brings us to the central concern of this volume. Since at least Poincaré, studies in various disciplines have been producing what Kuhn called anomalies, or cases left unexplained by the dominant science paradigm emphasizing the isolation of individual variables in closed experiments. Complexity and chaos were born as self-referential entities when the commonalities among these anomalies were recognized, and Kuhn's stage of 'revolutionary science' followed with the creative adoption of new approaches and the application of ideas to new disciplines. That was in the late 1980s and early 1990s, and is best exemplified by the formation of new institutes devoted to the study of nonlinear dynamics. Over the past 10 years, however, complexity has become more established and less *avant-garde*, and practitioners have focused more on Kuhn's 'normal science', working out the methodological aspects of how to model this or that phenomenon. The revolution is evidently over.

This book has its origins in our conviction that the revolutionary and exploratory period of nonlinear systems studies was too brief and incomplete, at least in the realm of the social sciences (see also Turner 1997). The CAS approach promulgated by SFI and other centres for complexity studies seems to imply a distinct theoretical package different from existing schools of thought in archaeology. The work by many in Europe on the other hand tends to suggest that various different theoretical orientations might profitably use nonlinear *tools* for their own questions and topics of research interest. These are quite different interpretations. This volume is intended to present the ideas of a variety of researchers known for their contributions to other areas of archaeology, with the hope that it may be possible to clarify what nonlinear systems signify for the discipline. Is it a new paradigm that offers nothing to those archaeologists who prefer studies of agency, hermeneutics, and multivocality, or does it cross-cut old divisions in a manner that can potentially unify the scattered remnants of archaeological thought left in the early 21st century?

It has recently become clear that practitioners sometimes have different ideas of what complexity is, and as a result adopt highly distinct forms of discourse. Some refer to it as a 'Post-Modern' science (Spencer-Wood 2000) because it eschews determinism and embraces contingency, while others see it as simply a development that unifies major trends within many of the sciences (see Price 1997). Some see nonlinear studies as a natural outgrowth of the computer revolution, allowing analysts to work with vastly increased volumes of data, while others see little need to adopt computer simulation to apply the theoretical principles involved. Our goal for this volume is to continue the revolution in the study of nonlinear systems by re-evaluating its theoretical bases and how they relate to paradigms already in place within the discipline. To this end we have assembled research from a broad variety of viewpoints as to what nonlinear dynamics might mean for archaeology. The articles in this volume are meant to bridge the gap between anthropological and archaeological theory and nonlinear concepts. In some chapters the authors have chosen to work with archaeological case studies to illustrate their points, but in others they have drawn upon ethnographic, historical, or contemporary societies where the characteristic ambiguities of archaeological data could be overcome. Regardless of the datasets used, the contributors are all concerned with the importance of nonlinear systems for archaeology.

The ability of humans to incorporate, integrate, and use information is the focus of Robert J. Hommon's investigation into the concepts behind CAS. Collective behaviour is presented as a result of 'schema-driven behaviour of interacting agents' where schema refers to internal, experience-based rules following Gell-Mann's use of the term (1992:10), but the definition could equally well have come from current authors in social theory. Hommon is interested in how human societies differ from other CAS by invoking processes of appropriatizing (encouraging conformity) and ecaptation (altering the environment), and the presence of stratified control hierarchies (unequal empowering of specific agents or groups of agents). Although primarily drawing upon examples from modern and recent human societies, he cites archaeological and historical evidence to demonstrate the influence of these three processes in Hawaiian culture prior to European contact.

Carole L. Crumley is also interested in human organization, but particularly those forms that do not correspond to the stereotypical stepped hierarchy. Since the interaction of components is one of the central elements of any complex systems approach, she focuses her chapter on those interactions, in particular the concept of heterarchy. A heterarchy is a meshwork of systems whose elements are either unranked or potentially ranked in a number of ways. The less rigid, more flexible structure of heterarchical systems is an adaptive tool that can more effectively combat 'surprise' within a society (e.g., environmental change, invasions, epidemics, etc.). Using examples of 'disorganized organizations' from recent times such as Al-Qaeda and the Anarchist movement, Crumley demonstrates that in uncertain times the sharing of information through flexible authority structures reduces risk by increasing available information to decision makers and multiplying solutions.

The fact that the concept of *agency*, in varying degrees, runs through many of these contributions reinforces the notion that the role of the individual is an important element of nonlinear studies. Many aspects of the nonlinear approach concentrate on mapping the emergent outcomes of the actions of individual agents, whether those agents represent people, ducks, or water molecules. Christopher S. Beekman builds upon prior suggestions by Nigel Gilbert that the authors of agent based simulations of human societies need to pay greater attention to the social actor and current social theory on agency. Prominent social theorists like Anthony Giddens, Margaret Archer, Pierre Bourdieu, and others have all discussed agency in different ways that have significant implications for agent based modelling. Beekman argues that, far from pursuing an empirical, 'bottom up' agenda with agent based simulation, modellers have unknowingly espoused particular theories about how the individual relates to society and left others unconsidered. He also argues that the empirical ideal that crops up repeatedly in enthusiastic portrayals of complex systems may not be possible. While many authors have emphasized self-similarity across scales (drawn from fractal analysis) as a central element of nonlinear studies, Beekman instead points to the work by Prigogine and by social theorists arguing instead for phase transitions, thresholds, emergence, and distinct scales of analysis with different rules. He finishes with an example drawn from Late Formative-Classic period (200 B.C.-A.D. 550) western Mexico to illustrate some of the complexities of collective agency.

Tammy Stone is also interested in the issue of quasi-group formation, or the appearance of highly unstable social groups in middle-range societies. Social institutions which aid in the flow and processing of information are argued to be self-organizing (Van der Leeuw 1981b; Stone 1999). Thus changes in information flow can have major repercussions on the structure of social groups which process information. As the intensity of flow among closely interacting groups increases, social complexity can differentially expand or collapse at spatially concentrated loci on the social landscape, forming information vortices (after Van der Leeuw 1981b). Stone applies these insights to quasi-group or faction formation as they occurred among the Hopi early in the 20th century. She contrasts the response at Orayvi against that of other Hopi communities that did not fission. Stone's discussion is consistent with Prigogine's argument that the current state of a system is one of many possible outcomes ultimately self determined by its initial conditions and by the response to subsequent disruptions in the flow of necessary resources (here, information).

William W. Baden examines the applicability of dissipative structures within archaeological models of agricultural systems. Baden provides an historical overview of anthropology's adoption of thermodynamic concepts, positioning Prigogine's nonlinear paradigm within anthropology's on-going theoretical use of these principles. By applying Prigogine's concept of systems far from equilibrium to prehistoric cultures, he updates and operationalizes Leslie White's (and his followers') original arguments linking the Second Law to cultural evolution. Using Mississippian examples from the southeast United States, Baden demonstrates how agriculture, as a dissipative structure, can be seen as an external source of negative entropy consistent with the theoretical suggestions of Schrödinger and Prigogine. Predictable disruptions in the flow of energy and their impact on social systems far from equilibrium are correlated with observed phase transitions in the archaeological record.

T.J. Wilkinson represents the MASS project and its work on settlement formation in northern Mesopotamia, and his chapter discusses the impact of simulation upon traditional landscape archaeology. Clearly influenced by Kohler's highly important simulations of the ancestral puebloans, the MASS project seeks to model an entire city of a complex, stratified society. This is a tall order, but is an excellent example of how agent based simulation is growing in ambition and capability. The subject has required increased attention to the spatial organization of subsistence and political economy. Wilkinson also clarifies how an agent based approach can improve upon traditional methods such as site catchment analysis. Instead of applying the fixed parameters of catchment analysis to a site, the MASS project models the activities of individual agents from below. While reading this article, one can easily envision future simulations that will model certain activities and predict the presence, volume, and diversity of material residue likely to result from those activities, providing a powerful tool for middle-range theory.

J. Stephen Lansing and Robert L. Axtell generously agreed to review the chapters in this book and give us their thoughts on the application of complex systems theory to archaeology. Lansing and Axtell have been at the centre of the maelstrom for over 15 years, and they discuss the merits of the expanding perspective from their unique vantage points in complexity research and agent based modelling.

Conclusion

The editors and contributors to this book do not seek to appropriate nonlinear studies for a specific area within archaeology. The diverse backgrounds of our contributors – Heterarchy, Social Agency, Information Theory, Factionalism, Systems Ecology, Landscape studies – should make that point clearly. Each has ties to one or the other of the theoretical lineages discussed above, and each shows how nonlinear dynamics might be considered within their own field. There are differing positions amongst the two editors as well, creating a dynamic tension that we find stimulating and fruitful. The differences amongst all contributors suggest that nonlinear systems models have provoked a curious realignment of theoretical traditions quite different from the outmoded Processualist vs. Post-Processualist divide. We hope that this volume will widen the field of debate over the utility of nonlinear systems for archaeology, and help other archaeologists evaluate the field differently than they might if they were only to consider the high profile work produced within some American institutions. Because, as Stone argues in her chapter, perception of the approach is as important as the approach itself.

We end our introduction with an anecdote to bring this point home. The editors originally organized the session that slowly and painfully inspired this volume for the 2002 American Anthropological Association meetings in New Orleans. At the time, there was a vocal 'pro-Science' splinter group that was holding its own sessions down the street from the conference hotel, sessions that had not been accepted into the official AAA programme because (according to the dissidents) they were too scientific. The group circulated a list of AAA sessions that did meet their criteria for a scientific approach, and our session was included. This unsolicited testimonial was bestowed on the basis of no more information than the title of the session – 'Nonlinear Systems Approaches in Archaeology' – all that was available to the group when they made their pronouncement. Our session participants showed mixed reactions to the news, and happily our audience represented a gratifyingly diverse cross-section of our anthropological colleagues. In the polarized aftermath of the debates within archaeology in the 1980s and 1990s, some archaeologists are willing to declare an approach valid or not on the most superficial of evidence. We direct this book to those members of the field who are dedicated to the work, and are still open to potentially effective and innovative ways of analyzing human society, regardless of what one calls them.

Chapter 2

As Water for Fish: Human Groups as Complex Ecaptive Systems

Robert J. Hommon
National Park Service

Introduction

Human behaviour differs radically from that of other organisms in three significant but inadequately explored ways that enable us to control other human individuals, our environments, and large social groups. Each of these three unique features depends on the cumulative nature of human culture, which is based in turn on the ability of our evolved cerebra to incorporate, integrate, and use enormous quantities of information.

The first of these three unique features, here called *appropriatizing*, is the ubiquitous interactive process, usually employing language, that enables one to affect a person's behaviour by triggering mental rules based on another's experience.

Second, in contrast to adaptation, which predominantly determines the relationship between all other species and their environments, human beings have developed *ecaptation*, the uniquely successful interactive process by which we alter environments to fit our needs.

The third unique feature of human behaviour is centrarchy, centralized control of behaviour by mass media and stratified control hierarchies. A stratified control hierarchy is a tiered structure which enables human beings to exercise centralized control in ways that differ fundamentally from the distributed or polyarchical control and simple hierarchical control evident in other animal groups.

Complex Adaptive Systems (CASs)

In recent years, a broad range of animate phenomena including organisms, cells, ecosystems, the human brain, herds of mammals, schools of fish, and flocks of birds, have been described as Complex Adaptive Systems (Gell-Mann 1992; Holland 1992, 1995). The term *system* in this context refers to a group of physical elements (such as animals or neurons) and their interrelationships; a *subsystem* of a system is understood to consist of some subset of these elements and their interrelationships. This *node and connector* definition of system contrasts sharply with the more abstract sense of the term in which the elements and subsystems are not quantifiable phenomena but categories such as the 'subsistence', 'religious', or 'political' functions of a 'cultural system'.

A *Complex Adaptive System*, or *CAS*, consists of interacting elements usually called actors or, as in this chapter, *agents*. For example, in a flock of birds considered as a CAS, each bird is an agent. The interactive behaviour of each agent in the system is determined largely by its internal, genome-plus-experience-based rules, here termed *schemas*. According to Gell-Mann (1992:10),

> The regularities of the experience are encapsulated in highly compressed form as a model or theory or schema. Such a schema is usually approximate, sometimes wrong, but it may be adaptive if it can make useful predictions including interpolation and extrapolation and sometimes generalization to situations very different from those previously encountered.
>
> In the presence of new information from the environment, the compressed schema unfolds to give prediction or behaviour or both.
>
> When the compression took place, regularities were abstracted from experience and compressed. The rest of experience, ascribable to chance or to regularities too subtle to recognize, cannot be compressed and does not typically form part of the schema. When the unfolding takes place, new material is adjoined, much of it again largely random, as 'present data' or input data from the real world.

Schemas of interest in a CAS are those that can be summarized as *if...then* or *condition-action* statements; *IF X* occurs, *THEN* response *Y* is appropriate (Holland 1995:7-8). For example, for a bird in a flock, one schema may be thought of as *IF* neighbouring birds increase their distance upward, *THEN* fly upward. Schemas are generated by the individual agent's:

- genome; and, often,
- direct experience of the environment; and, in some cases
- imitation of the behaviour of other agents.

Ideally, the simplest kind of schema, generated by genome alone, is an involuntary reflex, unmodified by experience or imitation. The individual agent's experience may, of course, affect the expression of many genetically-based schemas in a variety of ways. For example, a hawk's genetically-based predatory propensity can be enhanced by experience in both successful and unsuccessful attempts at catching prey. Experience in the broad sense can also include such factors as injuries that impair flying or walking, or malnourishment that arrests behavioural development. The third generator of schemas, imitation, appears to be weakly developed in most non-human animals.

An important defining feature of a CAS is its *mode of control*. The coherent behaviour of the CAS as a whole (its *global* behaviour) is usually not imposed centrally, as by a supervising agent, but is said to *emerge* from the interaction of the constituent agents, each behaving according to its own internal schemas (Epstein and Axtell 1996:33-35). *Polyarchy* is the term that is used here for this distributed or bottom-up control (Hommon 2000:136). The coherent behaviour of a flock of birds, for example, is not determined centrally, by some alpha bird, but rather is controlled polyarchically by all of the birds together, each behaving according to its internal schemas applied to current conditions. The nature of cognitive processes within an animal brain that generate and activate schemas is not yet well understood. However, the general notion of emergent behaviour finds support in simulations such as the computer software that can mimic coherent flocking behaviour among virtual birds obeying just three simple rules (Langton 1989:89; Reynolds 1987).

The behaviour of a CAS is *nonlinear*, which in this context means that the behaviour of the CAS as a whole is not necessarily proportional to the behaviour of any particular agent or subset of agents in the system. A single act by an agent may set off a cascade of interactions among agents that changes the state of the entire system. Conversely, altered behaviour of numerous agents may have little lasting effect on the system's behavioural trajectory.

Far more important than the magnitude of the initiating behaviour are the interconnectedness of the elements of the system and the state of the entire system at the time of the initiating behaviour. For example, in a flock of birds flying at sundown, a momentary loss in altitude of a single bird might trigger a general movement of the flock to a tree that would then serve as a roost for the night. Conversely, the flight trajectory of the same flock the following morning might be undisturbed if several birds lost altitude or left the flock entirely.

The term 'butterfly effect' is sometimes applied to large scale change in a nonlinear system triggered by an apparently minor variation in a small element in the system. The term originally referred to the behaviour of weather, described as a chaotic system. In principle, according to meteorologist Edward Lorenz, it is possible for the flap of a single butterfly's wings in Brazil to trigger a tornado in Texas. The notion is that if the atmosphere over Texas is on the verge of forming a tornado, that single butterfly flap could in theory set off a chain reaction that cascades through the system and eventually generates the tornado (Lorenz 1993:181-184).

Human Uniqueness

The nonlinear, adaptive system whose global behaviour emerges from the schema-driven behaviour of interacting agents is a powerful model for biological phenomena of many kinds at all scales. Human groups certainly exhibit all the defining features of CASs summarized above. In short, the global behaviour of human groups ranging in size from nuclear families to nation-states can be seen to emerge from the interaction of individual human beings, each of whom behaves according to internal schemas. However, I suggest that, in addition to this array of features, human beings also exhibit unique ways of interacting with each other and with their environments, and further, that these uniquely human behavioural processes are so fundamental and so pervasive that the CAS model as summarized here is insufficient to describe the dynamics of any human group.

A list of unique attributes of human culture might include aesthetics, conservation, cumulative learning, economic exchange, fiction, humour, intentionality, language, mathematics, mass media, philosophy, politics, recreation, religion, science, theory of mind, traditions, transportation, and warfare. Some might argue that such features are not literally unique to our species because each must have developed from behavioural precursors in pre-human ancestors and that similar characteristics are evident in contemporary non-human species. Further, it might be argued that these human behaviours and those of other species are merely situated at different positions along developmental continua, and therefore differ only in a quantitative sense.

In response, I suggest that certain *quantitative* gaps between human agent-agent and agent-environment interactions and such interactions exhibited by non-human species are so great that they are in effect *qualitative* differences. In other words, a difference in degree that is sufficiently extreme can be considered for present purposes a difference in kind. Think of a meter stick that is painted with a grey scale varying continuously from white at one end to black at the other. The white end can accurately be said to differ 'in degree' from the black end, yet the shades of the two ends if compared side-by-side will be seen to differ qualitatively.

We can apply our grey-scale meter stick metaphor to a comparison of culture among chimpanzees and human beings. A recent review article in the journal *Science*, summarizing a cumulative total of 151 years of research on wild chimpanzee groups by various scientists, listed 39 examples of learned, shared behaviour that the authors have identified as 'cultural' (Whiten, et al. 1999). While the discovery of a rudimentary chimp culture is important for an understanding of culture in apes, ourselves, and our ancestors, the chimpanzee cultural attainment virtually vanishes in quantitative terms when compared with the millions of cultural practices exhibited by *Homo sapiens*. Compare, for example, the preparation and use of a stalk of grass to catch termites, or the selection of stones to break open nuts with the construction and flight of the Space Shuttle, a piece of technology with 2,500,000 interacting parts.

The Space Shuttle illustrates well the cumulative nature of human culture, a feature that seems primarily attributable to the immense schema capacity of the human brain. With limited exceptions (such as the scant handful of chimpanzee traditions and the songs of species of birds and whales), we are the only animals whose non-genetically-based behavioural repertoire does not have to be created anew by every individual in every generation. We are spared the necessity of reinventing the wheel by our ability to accumulate and apply knowledge of wheels inherited from earlier generations. The roots of the techniques and physical components necessary to the Space Shuttle's construction, for example, can be traced back, through many stages, to the Bronze Age and even to the concept of standardized form that is evident in the ancient stone tool traditions. Such venerable cumulative pedigrees are undoubtedly common for virtually every element of culture.

The cumulative nature of human culture encompasses not simply an ability to pile up enormous quantities of schemas, of course, but also to compare, select, integrate, assemble, and combine them in seemingly limitless ways to generate new schemas, behaviour, and artefacts. Applying this ability, we and, to a limited extent, our progenitor species, have radically changed how we interact with each other and with our environments in three uniquely human ways that are described below: appropriatizing, ecaptation, and centrarchy.

Appropriatizing

As indicated above, the behaviours of most individual organisms in a CAS are generated by some combination of three factors: the individual's genome, direct experience of the environment, and, in some organisms, an ability to imitate the behaviour of others. For animals that interact frequently with others of their kind, such as our contemporary primate relatives, individual experience can include control exercised by members of the group by means of behaviour such as physical contact, gestures, facial expressions, and body postures. Similar behaviours among our pre-human ancestors were probably precursors of a fourth, uniquely human determinant of behaviour, a type of agent-to-agent interaction referred to here as 'appropriatizing'.

To *appropriatize* (or *prope*) is to act intentionally to affect a person's behaviour by instilling or activating a schema in his or her mind. It is the *intent* to change behaviour that defines proping. The level of success of a proping act depends on the reaction of the recipient, who is often resistant to change. Proping is pervasive in human interaction. Barring marked impairment, every human being beyond infancy has learned appropriate ways to act in a wide variety of circumstances, and is frequently engaged in trying to make others (and him- or herself) act accordingly. What is considered 'appropriate', of course, varies widely from group to group and individual to individual, and what is appropriate for one group, such as a teenage gang, is often inappropriate for another, such as a rival gang or the society at large.

In groups of interacting non-human animals such as our primate relatives, the mental state of an individual, expressed in behaviour such as slapping, tickling, biting, chasing or grimacing may trigger other animals' behaviour in response. In some respects interactive human behaviour may seem to resemble such actions among non-human species, but human agent-to-agent behaviour is far more frequent, nuanced, and complicated than that of any other animal. In part this human difference is attributable to our species' use of language and our ability to accumulate schemas in enormous quantities. Even with these abilities, human interactions would differ radically from those we experience if not for the uniquely human ability to 'mentalize'.

To *mentalize* is 'to understand and manipulate other people's behaviour in terms of their mental states' (Frith and Frith 1999:1692), an ability that appears to be absent in other species except possibly in rudimentary form in the great apes. Proping can be considered the active, manipulative element of mentalizing. Only we humans seem to be fully aware that others of our species are beings with minds and mental processes resembling our own. Using this insight we are able to form complex and often accurate beliefs about another's motives and to predict in some detail his or her future behaviour on that basis. By being able to imagine being in another's place (while sometimes adjusting for differences of sex, age, knowledge, and other factors), we can effectively encourage others to behave in ways we deem appropriate (proper, correct, right, moral, smart, cool, etc.) and discourage behaviour that we consider inappropriate (improper, incorrect, wrong, immoral, stupid, etc.).

Several lines of evidence support the contention that proping comprises a major element in human interaction. One indication of the pervasiveness of proping is the great variety of ways in which the proping process can be described. Appendix A is a compilation of more than 1,400 common English verbs that describe various ways of proping and responding to such behaviour. ('Common' means here that the verbs in the list might be found in a popular weekly news magazines such as *Time* or *Newsweek.*)

Literature, considered as a distillation or condensed version of experience, supplies a second body of evidence for the prominence of proping in human interaction. Analysis of certain well-known works of both classic and currently popular fiction, ranging from tragedy to light comedy yields high frequencies of proping behaviour.

For example, an audience viewing a full four-hour performance of Shakespeare's *Hamlet* (1992) will hear at least 805 instances, an average of one proping sentence every 18 seconds. In Miller's *Death of a Salesman* (1949) the 1,746 proping sentences are 3.5 times as frequent: once every 5.16 seconds in a two-and a-half hour performance of the play, and O'Neill's *Long Day's Journey Into Night* (1989) presents the audience with 2,157 proping sentences, or one per 3.8 seconds in a 136 minute performance. Nearer the light end of the fiction spectrum, are Rowling's phenomenally popular Harry Potter books, with over 175 million copies in print in more than 50 languages. *Harry Potter and the Sorcerer's Stone* (Rowling 1997), the first of the series, includes 2,024 proping sentences in 309 pages, an average of 6.55 per page. Twain's *Adventures of Huckleberry Finn* (1999), still a popular classic after more than a century and recognized today as one of the great American novels, exhibits 1,589 proping sentences in 284 pages, or an average of 5.6 propes per page. In the latter book, the only extensive sections that lack proping are those in which little or no human interaction is depicted. To expand the sample beyond 400 years of Anglophone culture, consider the *Odyssey* (Homer 1999). Composed in the 8th century B.C., this epic poem consists of 12,109 lines. The 1,277 proping sentences occur at an average interval of one every 9.5 lines, or one per 36.7 seconds in a 13-hour recitation. (The proping data in this case include interactions involving humans as well as gods and other supernatural beings exhibiting human-like behaviour.) I suggest that proping is not incidental to these works but is at the core of the emotionally-charged tension and conflict that has driven fiction for at least 2,800 years.

It seems evident that the popularity of a work of fiction depends in large part upon the degree to which readers or audiences are able to identify with the behaviour and emotions of the characters depicted in the work, that is we understand their behaviour in terms of their mental states, which resemble our own. No matter how unfamiliar the setting and events of a work of fiction, characters are usually expected to behave in ways that are somewhat consistent with people's actual experience. Accordingly, the great popularity of the works of fiction listed above indicates that the high frequency of proping depicted therein is considered by readers and audiences to resemble actual human interaction.

The prevalence of proping is also evident in humour. I suggest that a reasonable definition of laughter, in most contexts, is the barking sound that human beings make in the presence of the inappropriate (Hommon 2001:143). Perhaps the barking of our early forebears at those guilty of behaviour deemed inappropriate served to control deviance much as the laughter accompanying ridicule and gossip do today.

The importance of proping in humour can be supported by a perusal of the comics section of any newspaper on any given day, where more than half the strips are likely to include an instance of proping. Proping often generated the humour of Charles Schultz's comic strip 'Peanuts', which, according to Lyman (1999), has been the most popular comic strip in the world, in terms of breadth of distribution (2,600 newspapers in 75 countries and 21 languages). For example, in the recently published first volume collection of 'Peanuts' (Schulz 2004), proping figures in 65% of the 312 strips of the first year (1950-51). The *New Yorker* magazine, widely recognized for the excellence of its cartoons, provides a more recent example. Analysis of the cartoons published in the 23 issues of the first half of 2003, for example, shows that 68% of the 475 cartoons were based on proping.

The examples presented here support the notion that proping permeates human interaction. I suggest that the human capacity for language has evolved and currently functions not only simply to convey information but more specifically to apply and respond to proping in the variety of ways indicated in Appendix A. Further, I suggest that proping is the primary process by which we humans transfer most of the schemas that generate cultural behaviour.

By turns endlessly innovative and deeply conservative, the human capacity for proping both liberates and confines. At the most fundamental level, the ability to prope allows humans to base behaviour on information sources who are indefinitely distant in time, space, and number of intermediaries, rather than having to depend on schemas generated by each individual's genome and direct experience. For each individual in a human group the proping process provides novelty in the form of schemas to be incorporated during enculturation. From the perspective of the human group on the other hand, proping tends to be a highly effective conservative process that ensures continuity of the group's existing traditions and customary behaviour. Probably our ancestors were selected in part for their ability to give and accept proping such as advice that precluded the necessity of learning about large bears and other threats first hand.

From time to time, significant numbers of agents in a human group choose to favour variant forms of schemas that begin in the minority, leading to what might be called a social butterfly effect, and eventually, in some cases, the dominance of the new schema and the extinction of the old. If a shift in gene frequencies in a breeding population is a definition of biological evolution, then we can describe such a shift in schema frequencies in a human group as an example of cultural evolution.

Behaviour about which humans are willing to prope run the gamut from how to hammer a nail to intricate compliance with codes of morals, ethics, and etiquette. Proping interactions almost always include emotional elements, employed in the service of influencing the behaviour of others. The graphic, plastic, and performance arts represent a kind of interactive behaviour, which we can call *aesthetizing*, whose primary intention is to control not so much peoples' behaviour, as their emotions.

Ecaptation

The second uniquely human process, *ecaptation*, is the opposite of evolutionary adaptation. The term *adaptation* can refer to widely varying time-scales, morphology, and behaviour, from rapid reactions of organs to stimuli, to changes within an individual organism's lifetime, to species' trajectories across thousands or millions of years. By means of *evolutionary adaptation*, the process discussed here, a species or breeding population changes in ways that increase its fitness to its environment (Rudin 1997:5).

While our species obviously shares with our ancestors at all taxonomic levels a broad range of physical and behavioural adaptations, we also employ an extremely effective process, here called *ecaptation*, which reverses the direction of the adaptation process. The term ecaptation is derived from the prefix 'ec-' (as in ecology), originally from Greek *oîkos*, 'house', and Latin *aptāre*, 'to fit'. We can define ecaptation as the process by which an organism intentionally alters its environment for the purpose of increasing its fitness to the organism's perceived needs. Note that 'fitness' in this context does not refer to Darwinian suitability of a species to an environment as demonstrated by the species' reproductive success but rather how well the environment satisfies the perceived requirements of the organism.

The concept of adaptation stresses the control that the environment exercises over the organism. In contrast, ecaptation emphasizes the control that the organism exercises over its environment. Adaptation is to natural selection as ecaptation is to ecaptive selection. In natural selection, the environment 'chooses' certain variants in an organism's form or behaviour, thereby increasing the organism's fitness to the environment. In contrast, in ecaptive selection, the organism chooses and alters portions of the environment to increase their fitness to the organism.

Many non-human species are capable of some degree of ecaptation, of course. A wide variety of species of invertebrates, fish, reptiles, birds and mammals build nests and burrows, for example, and a few animals make and use simple tools. Human beings, however, are the ecaptive species *par excellence* because the schemas that determine the ways we alter environments to serve our purposes are generated not only by genome, direct experience and imitation, as in other species, but also by proping. Proping and our enormous schema capacity allow individuals in each human generation to incorporate the knowledge of past generations and to add to such knowledge before passing it on. It is this cumulative ability that sets human ecaptation apart from that of all other species. One rough quantitative measure that can be applied to the uniqueness of human ecaptation is the number of steps necessary to create an artefact from naturally occurring resources. The ecaptive steps required for the 2.5 million interacting parts of the Space Shuttle, for the moment a pinnacle of human technology, probably exceeds those exhibited by non-human species and our earliest hominine ancestors by more than seven orders of magnitude. While other animals build nests and occasionally collect food with the aid of a stick or stone, only human beings have transformed the world.

We and some of our ancestral species have been modifying the environment at all scales so long and so intensively that we have created what can be termed the *factosphere* (from the same root as 'artefact'), which is virtually as significant for our species as are the biosphere, geosphere, and atmosphere. The factosphere consists of all our artefacts in the broadest sense: physical phenomena that have been transformed or transported from natural forms or locations by deliberate human action. Components of the factosphere include not only the more obvious elements such as screwdrivers, cathedrals, the Mona Lisa, and the city of Tokyo, but also cultivated lands, compressed air, whiskey, ice cubes, domesticated animals, and almost every bite of food we eat.

The global scale and complexity of our ecaptive activities are indicated by the following observations. The estimated 45 gigatons of soil and rock moved by humans every year exceeds by at least six gigatons the amount moved by all the world's rivers (Hooke 1994; Houtman 2004). Hooke (2000:845) estimates that within the next century, humans will move enough earth to build a mountain range 100 kilometres long, 40 kilometres wide and 4,000 meters high, a volume equal to what we have moved in the last 5,000 years. According to Imhoff, et al. (2004) humans currently incorporate roughly one-third of the planet's yearly net primary production, that is 'the net amount of solar energy converted to plant organic matter through photosynthesis'. This estimate is about 60 times what might be expected given the fact that our species constitutes only 0.5% of the planet's heterotrophic biomass. Of equal significance, though at much smaller scales, we have taught ourselves to rearrange individual atoms, produce antimatter in quantity, build viruses, and create transgenic organisms.

Our ancestors have been ecapting environments for at least 2.5 million years (Klein 1999:217), first by transporting and modifying portable objects such as sticks and stones; later by managing fire, weaving nets and baskets, building artificial shelters, trading goods, domesticating organisms, building cities, and inventing all the current technologies that flesh is heir to. The archaeological record of early long-lived stone tool traditions giving way to briefer and more variable traditions indicates that about 250,000 years ago our ancestors' ability to accumulate innovations began to accelerate (Klein 1999:338-341). The fact that today proping in various sophisticated forms (teaching, training) is essential to passing on multi-generational ecaptive traditions of all kinds, from plumbing to gene-splicing, suggests that advances in ecaptation in the past accelerated as our ancestors evolved and developed their proping skills.

Since their earliest appearance in the archaeological record, artefacts have been superior to what natural selection provided and could ever provide to our ancestors. Our tools have been sharper and harder than our teeth and nails; our baskets and container ships can carry far more than our arms alone; our agricultural fields, herds, and factories concentrate perennial supplies of food far more effectively than nature ever could. Our machines allow us to travel much farther and faster, to see and hear much more acutely, to communicate more broadly, to live longer, and to store far more information, than would ever be possible without ecaptation. In short, ecaptation has allowed us to transcend a wide range of physiological and behavioural limits imposed by evolution.

The ecaptively-generated factosphere permeates human culture. Today more than 2 billion members of our species live in vast cities, anthropogenic environments where the only significant environmental components that have not been deliberately transformed, transported or processed may be the air and solar radiation. Much the same can be said of many occupants of rural areas, though the degree of transformation of the soil of pastures and agricultural fields, for example, is not as profound as that of buildings of steel, concrete and plastic.

Ecaptation enables us to live virtually anyplace on land or at sea, and to survive for months at a time under water or in outer space. Without ecaptation, we, like our chimpanzee cousins, would never have extended our range much beyond central Africa, where limits on naturally occurring shelter, warmth, food, and water would have ended attempts at colonization. We have expanded our range not by means of adaptation to newly encountered environments but by altering portions of those environments to suit ourselves.

The natural conditions capable of selecting against any human breeding population that lacks the appropriate ecaptive array (for example fire, shelter, clothing, and weapons in the Arctic) still exist. For the most part, ecaptation does not *eliminate* the large-scale natural conditions such as polar cold or desert aridity that exert such selective pressures. Rather, our ecapted environments shield us from these conditions and in so doing, tend to nullify their adverse effects. To ensure simple survival and to pursue the great variety of human endeavours we create capsules or bubbles of ecapted environments that exclude or alter natural conditions (heat, cold, food scarcity, etc.) that would otherwise prove inconvenient, dangerous, or lethal. Broadly speaking, these ecapted environments can vary in size from a field of herded atoms in an electron microscope to vast urban complexes and continent-spanning agricultural landscapes.

According to one recent estimate, taken together, our ecaptive bubbles now comprise about 40% of Earth's land surface or about 60 million square kilometres (32 million square miles), or the equivalent of the entire Eurasian continent (National Geographic 2002). These anthropogenic landscapes include about 11% crop land, 25% pastureland, and 4% urban land (Chen 1990) and are growing rapidly in areas such as the tropical rainforests, which are being demolished at an annual rate estimated at 60,000 square kilometres (Willis, et al. 2004). Recent research in rainforests of South America, Africa, and Southeast Asia has shown that anthropogenesis predates the modern era by as much as 8,000 years in these areas once commonly thought of as largely 'pristine' or 'undisturbed' (Denevan 1992; Willis, et al. 2004).

An alternative to viewing artefacts as elements of the ecapted environment is to consider them to be parts of the extended human phenotype. Dawkins (1982:292) introduced the concept of the *extended phenotype* in the book of the same title, where he defined phenotype as follows:

> The manifested attributes of an organism, the joint product of its genes and their environment during ontogeny. A gene may be said to have phenotypic expression in, say, eye colour. In this book the concept of phenotype is extended to include functionally important consequences of gene differences, outside the bodies in which the genes sit.

An example of the extended phenotype to which Dawkins returns several times is the beaver dam.

Adapting Dawkins' concept to culture in general and archaeological data in particular, Dunnell (1989:45) has asserted that '[a]rtifacts do not "represent" or "reflect" something else that is amenable to evolutionary theory; they *are* part of the human phenotype'. O'Brien and Holland (1992:37) share this view. Recognizing that the term 'artefact' is not limited to portable objects, if we accept Dunnell's definition, then we must ask how far a human being's phenotype extends. Dawkins (1982:233-234) himself, using the example of a beaver dam and pond, addresses the question of the size of his 'extended phenotypes' as follows:

> And how far afield can the phenotype extend? Is there any limit to action at a distance, a sharp cut-off, an inverse square law? The farthest action at a distance I can think of is a matter of several miles, the distance separating the extreme margins of a beaver lake from the genes for whose survival it is an adaptation. ... [B]eyond a certain size of beaver lakes, it would become hard to regard further increases in size as adaptations. The reason is that, beyond a certain size, other beavers than the builders of the dam are just as likely to benefit from each increase in size as the dam builders themselves. A big lake benefits all the beavers in the area, whether they created it or whether they just found it and exploited it.

It seems clear that, in Dawkins' view, the extension of an individual animal's phenotype is limited by that animal's individual behaviour, which is based, as we have seen in the case of organisms in general, on the individual's own genome, his or her individual experience, and (perhaps) imitation of others' behaviour.

Accepting Dawkins' notion for the moment, if a human being, we'll call her Sue Generis, invents a wholly new artefact, entirely unassociated with any pre-existing artefact tradition, then that artefact is part of Sue's extended phenotype, as defined by Dawkins. On the other hand, if Sue knowingly makes an artefact that is based on a pre-existing artefact or design, then it is not part of her individual phenotype, rather, if it can be said to be anyone's extended phenotype, then it is that of its original inventor. The same principle applies even more emphatically to an artefact that Sue uses, but did not make. Is the can-opener that Sue buys part of her phenotype? No, in the same sense that a prairie dog hole that a rattlesnake moves into is not in any useful sense part of the snake's phenotype. Following Dawkins' caution concerning beaver dams, and given culture's cumulative nature and the great time-depth of virtually all kinds of artefacts, it appears that artefacts cannot be considered the phenotype of any living person.

One approach to this problem might be to consider the factosphere to be the extended phenotype not of individuals, but of an entire species or breeding population. In this case, our species' phenotype would include, for example, the 60 million square kilometres of Earth's ecapted environment and the whole Space Shuttle fleet. If this were the case, then all the beaver dams in Oregon would have to be considered parts of the phenotype of a beaver that spent its entire life in Wyoming, clearly an untenable position. More importantly, biologists, Dawkins included, use the term phenotype to refer to individual organisms, not entire species (Dawkins 1982:292; Mayr 2000:289; Ridley 1996:671; Rudin 1997:284). Neither the individual nor the species application of Dawkin's concept to *Homo sapiens*' cumulative culture seems likely to shed light on any significant questions about the biological or cultural nature of human beings. Instead, the failure of this over-extension of the phenotype concept serves to emphasize the uniqueness of human culture.

Stratified Control Hierarchy

We and our hominine relatives have been ecapting environments for at least 2.5 million years. Proping probably evolved along with *Homo sapiens*' facility with language, sometime between 200,000 and 50,000 years ago (Carroll 2003; Enard, et al. 2002). Of much more recent origin is *centrarchy*, the third uniquely human process considered here. Centrarchy, the centralized control of the behaviour of large groups of individuals, is applied by means of stratified control hierarchies (discussed in more detail below) and mass media, both of which probably first appeared approximately 6,000 years ago. Phases of mass media development have been marked by the invention of writing, the alphabet, printing with moveable type, and modern electronic media.

Like ecaptation and proping, centrarchy appears to have developed from, yet differs significantly from, superficially similar control processes in other organisms. One of the fundamental principles of complexity theory is distributed control (polyarchy). Though insight into emergent processes in CASs can be gained from computer simulations in which all agents' capabilities and schemas are identical, living agents in biological CASs usually differ from each other in both variables and such variants can alter the behaviour of the system as a whole. A bird these with an injured wing, for example, may shift the flight trajectory of its flock.

An additional step away from homogeneous polyarchical control is exemplified in some species by simple dominance hierarchies in which certain agents exercise control asymmetrically over others in the group.

Simple hierarchies that resemble those in social groups of non-human animals are common in all human societies today. Control of family activities, for example, tends to be exercised by parents and other family elders. The uniquely human invention called here the stratified control hierarchy undoubtedly developed from such simple hierarchies, but differs from them in form and function.

A *stratified control hierarchy* is a sub-system of a human group consisting of a central agent (or small cluster of agents) that directs task-oriented behaviour of other agents who are organized in multiple tiers that increase in population and diminish in span of responsibility with increasing remoteness from the central agent. Such sub-systems form the familiar 'top-down' organizational pyramids found in nation-states, corporations, armed forces and other large, task-oriented human groups. Typically, orders and sanctions are distributed outward through the expanding tiers (i.e., downward through the organizational 'pyramid') and, in response, services, information, and goods (such as taxes) are sent toward the centre by or under the control of the agents in the hierarchy. Military organizations generally represent the most rigidly organized cases of hierarchical organization. The U.S. Army, for example, is organized in nine strata or tiers, including the Army as a whole, field army, corps, division, brigade, battalion, company, platoon, and squad. Each unit of each tier is led by a person of stratum-appropriate rank who commands subordinates and responds to commands of superiors (Bluhm and Motley 1995:214).

Stratified control hierarchies have proliferated in many areas of life and are essential to the operation of nation-states and other large-scale human groups. However, schemas that underlie customs and traditions that largely determine peoples' lives still tend to be maintained polyarchically.

In spite of the fact that nation-states are typically called 'complex societies', the stratified control hierarchies around which they are constructed do not operate by mimicking polyarchy or by replacing it, but rather by vastly *simplifying* the process of control required for large-scale tasks. For example, consider that polyarchical control of a small, isolated community consisting of 200 people can require as many as 19,900 dyadic interactive connections, each of which is constantly guided by the myriad schemas that compose the culture of the group. While polyarchical control is demonstrably capable of maintaining the cohesion of human social groups, it tends to be poorly suited to organizing large numbers of people to accomplish even moderately complicated tasks. The stratified control hierarchy avoids the complexity problem by dividing each task into numerous layered sub-tasks to be accomplished by people at various levels of the hierarchy. This division of labour sharply reduces the number of dyadic relationships that have to be maintained by each participating individual (Johnson 1982; Hommon 1996, 2000, 2001). Not only are the required interactions reduced to those between each agent and the agent's supervisor and small group of subordinates (if any), but also the number and range of schemas specific to the task (i.e., orders) tend to be relatively few and simple, especially when compared to the schemasphere of the social group as a whole.

In the span of a few thousand years, polities employing stratified control hierarchies have established sovereignty over virtually all of Earth's land and adjacent marine resources (even much of Antarctica has been claimed). Modern nation-states are as much as a million times the size of the largest human groups that existed as recently as 6,000 years ago. Wars among nation-states are fought by armed forces organized as stratified control hierarchies. In recent decades enormous corporations, organized on the same basic principles have come to control the world's business. The world would be vastly different but for this third uniquely human feature, centrarchy.

Ecaptive Complex Systems (ECSs)

As we have seen, human groups are complex systems that share certain significant features with all Complex Adaptive Systems (CASs). Like other organisms, human beings behave according to schemas generated by genome, direct experience, and imitation; are capable of adaptation; and exercise polyarchical control. However, human groups also exhibit three unique processes that operate in ways that are contrary to those of non-human CASs. We apply proping to control individuals, ecaptation to control our environments, and centrarchy to control large groups of people.

These three processes so pervasively and definitively determine human behaviour and so sharply distinguish human social groups from those of all other species that it is useful to describe human groups as Ecaptive Complex Systems (or ECSs) rather than as CASs.

The Study of Ecaptive Complex Systems: Hawaiian Examples

The following examples apply the ECS perspective to archaeological and ethnohistoric data on Ancient Hawaii (Hommon 1976, 1986, 2000, 2001).

The early Polynesian colonists in the Hawaiian Islands depended heavily on their domesticated animals and plants, as well as the full range of tools, skills and knowledge that they brought from their home islands as they immediately set about transforming environments on a large scale, beginning with the construction of irrigated pond-field systems for taro in lowlands and valley bottoms.

In most regions of the geologically young Hawaiian Islands, soils are relatively shallow and rocky, a fact that has benefited archaeology greatly. In the process of bringing the land under cultivation and other activities, countless stones had to be tossed out of the way, dumped in piles and rows, built into retaining walls, and otherwise moved or disposed of. As a result, archaeologists are often faced with what are often referred to as 'wall-to-wall sites', intact material evidence extending over hundreds of hectares of continuous ecapted landscape.

The rise of powerful kingdoms in the 18th century depended on the dry-land crops grown in the Kona and Kohala field systems on the west side of Hawaii Island, which together comprised about 20 thousand hectares of extensively transformed and intensively managed agricultural lands.

Hawaiians managed undomesticated fish species, primarily mullet and milkfish in some 360 artificial fishponds that ranged up to 86 hectares in size and together comprised about 2,300 hectares.

The indigenous Hawaiian kingdoms were composed mainly of commoner (*maka`ainana*) and chiefly (*ali`i*) classes. The lives and interests of commoners were almost entirely limited to the local community land unit, the *ahupua`a.* Hawaii's roughly one thousand *ahupua`a* typically extended inland from inshore waters so that they contained all or most of the necessities of life, from marine resources, salt, and coastal habitation areas, to inland agricultural zones and upland forests. Each *ahupua`a* community exercised polyarchical control over everyday matters of agriculture, fishing, craftwork, community ritual, and socializing within its boundaries.

In sharp contrast to the commoners, Hawaiian chiefs were organized in elaborate, stratified control hierarchies that were based both on genealogical status and political power. Linguistic evidence suggests that early in Hawaiian history chiefs were considered by commoners to be senior relatives within corporate kinship units. By the time of Western contact, commoners no longer belonged to such units and were forbidden to maintain genealogies (Kamakau 1992:242), with the result that they could no longer reckon genealogical connections with chiefs. The disintegration of kinship bonds facilitated the formation of polities based on conquest of multi-community districts and entire islands, because conquered commoners were not bound by kinship to the defeated chiefs, and were thus less likely to rebel against the newly imposed government. The process by which ancient Hawaiian polities based on non-egalitarian kinship units ('conical clans') became polities based largely on political power is not yet well understood, but genealogies and traditional histories seem to indicate that the transition may have been rapid and may have happened in the early 17th century, a time of intense competition for resources (Hommon 1976, 1986).

The indigenous kingdom on the island of Hawaii encountered by Captain Cook in 1779 encompassed roughly 10,000 square kilometres, with a population of about 100,000, which is well above the upper limit of middle-range societies as defined by Feinman and Neitzel (1984). The population of the Hawaii kingdom was similar in estimated size and density to early so-called 'complex societies' of the Tigris-Euphrates lowlands in the Old Babylonian Period, First Dynasty Egypt, the Basin of Mexico in the Toltec period, or the Central Maya Lowlands in the Middle to Late Pre-Classic or Early Post-Classic (Hommon 2001; Whitmore, et al. 1990). Investigating the origin of Hawaii's indigenous kingdoms will contribute significantly to an understanding of the emergence of large-scale society in part because Hawaii's dense population, agricultural economy, and other factors resemble those of other pristine archaic states. More important, however, are those factors common in such states elsewhere in the world that were absent or poorly developed in Hawaii, including long-distance trade, markets, permanent capitals, towns and cities, palaces, large-scale storage facilities, standing armies, draft animals, woven cloth, monetary systems, writing, metallurgy, and pottery (Hommon 2001). The Hawaiian example supports the view that none of these features are required for the emergence of large-scale societies. On the other hand, the fact that the governmental structure of the Hawaiian kingdom, like those of better known states, was organized as a stratified control hierarchy supports the contention that such subsystems are essential to the emergence and functioning of large-scale societies.

Summary and Conclusion

The Ecaptive Complex System perspective outlined above emphasizes human uniqueness by comparing three ways in which we exert control in the world with the behaviour of other species.

- In addition to the genetic, experiential, and imitative generators of behaviour available to other organisms, only humans depend markedly on *proping,* which allows us to accumulate vast stores of behavioural rules (schemas) based on the experience of others.
- In contrast to other organisms, whose survival is dependent on the ability to adapt to effective environments, humans are distinguished by their extraordinary ability to *ecapt*, that is, to intentionally alter environments for the purpose of increasing the fitness of those environments to our perceived needs.
- Only human beings have developed stratified control hierarchies and mass media which accomplish large-scale tasks by imposing strongly centralized control in contrast to the distributed, polyarchical control that is common in Complex Adaptive Systems at all scales.

For the moment, the success bequeathed by these three human innovations in a biological sense is manifest by our sheer numbers and global distribution. At the same time, flaws inherent in each of the three processes can lead to negative results. For example, clear-cutting a forest (ecaptation), an act enshrined by tradition as practiced by revered ancestors (proping), and required by government order (centrarchy) might lead to the collapse of an ecosystem that is required for human survival. No other organism would be capable of comprehending such a process or of perpetrating such folly.

The three interconnected human processes discussed here are probably rooted in ancestral abilities common to many species such as learning from experience, imitating the behaviour of others, constructing nests, and participating in simple dominance hierarchies. Such historical continua are deceptive, however. In much the same way that living things differ qualitatively from the non-living chemicals from which they evolved, human beings are not simply another species of animal. The three processes have enabled humans to transcend limits that would otherwise be imposed by our biological nature: freed of the requirement for direct interaction, we can apply knowledge gained at many removes from those distant in time and space; we are constantly involved in deliberately transforming, rather than being transformed by our environments; and we cooperate in large numbers to accomplish complicated tasks uncomprehended by other species.

Perhaps we have only incompletely appreciated the significance of these uniquely human features not because they are rare, obscure, or difficult to decipher, but because they are extremely common and obvious; as enveloping, essential, supportive, ubiquitous, and therefore as transparent and ignored as water is for fish.

Acknowledgements – The opinions expressed in this chapter are those of the author and are not to be construed as official or reflecting the views of the United States Department of the Interior.

APPENDIX A. 1,416 VERBS PERTAINING TO APPROPRIATIZING

Abandon; Abide by; Absolve; Abuse; Accede; Accept; Acclaim; Accommodate; Accompany; Acculturate; Accuse; Accustom; Achieve; Acknowledge; Acquaint; Acquiesce in; Act; Act out; Act up; Adapt to; Address; Adhere; Adjust to; Admire; Admit; Admonish; Adopt; Adore; Adorn; Advise; Advocate; Affect; Affirm; Afflict; Affront; Aggravate; Agitate; Agonize; Agree; Aid; Alarm; Allay; Allow; Ally; Alter; Amaze; Amuse; Analyze; Anger; Annoy; Answer back; Anticipate; Ape; Apologize; Appall; Appeal; Appease; Applaud; Appraise; Appreciate; Apprise; Approach; Approve; Arbitrate; Argue; Arouse; Arrange; Ask; Aspire; Assail; Assent; Assert; Assess; Assign; Assimilate; Associate with; Assume; Assure; Astonish; Astound; Atone; Attack; Attain; Attempt; Attract; Attribute; Avenge; Avert; Avoid; Award; Awe;

Baby; Back; Back off; Backbite; Back down; Backslide; Badger; Badmouth; Baffle; Bait; Balk; Ban; Banish; Bar; Bargain; Bash; Battle; Bawl out; Bear down on; Bear with; Beseech; Beat; Beat up; Beautify; Bedevil; Befriend; Beg; Begrudge; Behave; Behoove; Believe; Belittle; Bend; Berate; Besmirch; Best; Bestow; Bet; Betray; Better; Bewilder; Bewitch; Bias; Bicker; Bitch; Blackball; Blacken; Blacklist; Blame; Blast; Blather; Blend in; Bless; Block; Blunder; Bluster; Boast; Bond with; Boo; Boost; Bootlick; Bore; Boss; Botch; Bother, Bow; Brag; Brainwash; Brand; Brawl; Breach; Break (rules, people); Break down; Break in; Bribe; Bring down; Bring up; Broaden; Browbeat; Brown-nose; Brush off; Buckle down; Buckle under; Budge; Bug; Bulldoze; Bullshit; Bully; Bungle; Buoy up; Butter up; Buzz off;

Cajole; Calculate; Call; Call for; Call on; Calm; Capitulate; Captivate; Care; Care for; Caricature; Carp; Cast out; Castigate; Catch; Catch on; Catch up; Categorize; Cause; Caution; Cave in; Celebrate; Censor; Censure; Challenge; Change; Characterize; Charm; Chasten; Chastise; Chat; Chatter; Cheat; Check out; Cheer for; Cheer up; Cherish; Chew out; Chicken out; Chide; Chill out; Choke; Choke off; Choose; Citify; Civilize; Classify; Clean up; Clear up; Climb; Cling to; Clown; Cluck; Coach; Coax; Coddle; Coerce; Collaborate; Comfort; Command; Commend; Comment; Commiserate; Commit; Communicate; Compare; Compel; Compensate; Compete; Complain; Compliment; Comply; Compromise; Conceal; Concede; Concern; Conclude; Concur; Condemn; Condescend; Condone; Confess; Confide; Confirm; Conflict; Conform; Confound; Confront; Confuse; Congratulate; Connect with; Connive; Consent; Consider; Console; Conspire; Constrain; Construe; Contend; Contest; Contradict; Contrive; Control; Convert; Convince; Cool; Cool down; Cooperate; Cope with; Cop out; Copy; Correct; Correspond with; Counsel; Count on; Counter; Cover for; Cover up; Cow; Cower; Crab; Credit; Cringe; Criticize; Cross; Crow; Crush; Cry; Curb; Curse; Cuss at; Cut; Cut off;

Damn; Dare; Darn; Date; Daunt; Deal with; Debate; Deceive; Decide; Declaim; Decry; Defame; Defect; Defend; Defer to; Defile; Deflate; Deflect (as criticism); Defy; Degrade; Dehumanize; Deign; Delay; Delight; Delude; Demand; Demean; Demonize; Demonstrate; Denigrate; Denounce; Deny; Deodorize;

Depart; Depend; Depict; Deplore; Deprecate; Depreciate; Depress; Deride; Derogate; Describe; Desecrate; Desert; Deserve; Desist; Despair; Despise; Detect; Deter; Determine; Detract; Develop; Deviate; Dictate; Differ; Dig; Dignify; Direct; Dirty; Dis; Disassociate; Disagree with; Disappoint; Disapprove; Discipline; Disclose; Disconcert; Discount; Discourage; Discover; Discredit; Discriminate; Discuss; Disdain; Disenchant; Disgrace; Disguise; Disgust; Dishonour; Dish; Dish out; Disillusion; Disinherit; Disinvite; Dislike; Dismay; Dismiss; Disobey; Disown; Disparage; Displease; Dispute; Disregard; Disrespect; Dissatisfy; Dissent; Dissuade; Distinguish; Distress; Distribute; Distrust; Disturb; Diverge; Divide; Do (right, wrong, etc.); Dog; Domesticate; Dominate; Dote on; Double-cross; Doubt; Downplay; Dread; Dress down; Drift; Drink to; Drive; Drum in; Drum out; Dump; Dump on; Dwell on;

Earn; Ease; Ease up on; Echo; Edify; Educate; Egg on; Eject; Elbow; Elevate; Embarrass; Embody; Embrace; Empathize; Empower; Emulate; Enchant; Encourage; Encroach; Endanger; Endorse; Endure; Enforce; Enjoy; Enlighten; Enrage; Ensnare; Enthrall; Entice; Entrap; Envy; Err; Escape; Eulogize; Evade; Evaluate; Even (a score); Exaggerate; Exasperate; Exchange; Excite; Exclude; Excommunicate; Excuse; Exhort; Exile; Exonerate; Expect; Expel; Explain; Expose; Extol; Extrapolate; Eye;

Face; Face down; Face up to; Facilitate; Fail; Fake; Fall; Fall out; Falter; Familiarize with; Fascinate; Fault; Favour; Fawn; Fear; Feel for; Feign; Feud; Fidget; Fight; Figure out; Find Fault with; Fit in; Fix; Flame; Flatter; Flaunt; Flim-flam; Flip off; Flip out; Flock; Floor; Flout; Flub; Flummox; Fluster; Follow; Fool; Forbear; Forbid; Force; Forgive; Forgo; Forsake; Foster; Foul up; Fret; Frighten; Frown at; Frown on; Frustrate; Fulfill; Fumble; Fuss;

Gag; Gall; Gamble; Gape; Gasp; Gather; Gawk; Gesture at; Get (it); Get along; Get around; Get away; Get away with; Get back at; Get behind; Get even with; Get real; Get through to; Get to; Get with; Gibe; Give; Give in; Give up; Give way; Glare at; Gloat; Glorify; Gloss over; Goad; Go ahead; Go along; Goof; Gossip; Grab; Grade; Grant; Grasp; Grate on; Gratify; Greet; Gripe; Groan; Groom; Gross Out; Grouch; Ground; Grouse; Grovel; Grumble; Guard; Guide; Gush;

Hail; Halt; Hammer; Hamper; Handle; Hang (out) with; Harangue; Harass; Harmonize; Harp; Harry; Hassle; Hate; Haze; Hear; Heckle; Heed; Help; Henpeck; Hesitate; Hex; Hide; Hide out; Hinder; Hint; Hiss; Hit; Hit Back; Hold; Hold Back; Hold forth; Hold up; Hole up; Honk at; Honour; Hoodwink; Hope; Horrify; Hound; Hug; Humanize; Humble; Humiliate; Humour; Hunker down; Hurry; Hurt; Hush; Hustle; Hype;

Identify as; Identify with; Idolize; Ignore; Imagine; Imbue; Imitate; Impart; Impede; Impel; Impersonate; Impinge; Implicate; Implore; Imply; Impose; Impress; Improve; Impugn; Impute; Incense; Incite; Include; Inculcate; Incur; Indoctrinate; Indulge; Induce; Infer; Inflame; Inflate; Inflict; Influence; Inform; Inform on; Infuriate; Infuse; Ingrain; Ingratiate; Inhibit; Initiate; Injure; Insinuate; Insist; Inspect; Inspire; Instigate; Instill; Instruct; Insult; Intend; Interact; Intercede; Interchange; Interest; Interfere; Internalize; Interrupt; Intervene; Intimidate; Intrude; Inveigh against; Invite; Irk; Iron out; Irritate; Isolate;

Jar; Jawbone; Jeer at; Jeopardize; Jilt; Jitter; Join; Joke about; Josh; Jostle; Judge; Jump on; Justify;

Keep away; Keep out; Keep up; Kibitz; Kick; Kid; Kill; Kiss; Knock; Know; Kowtow to; Kvetch;

Label; Lag behind; Lament; Lampoon; Lash out; Laugh at; Laugh off; Laugh with; Lay low; Lay off; Lead; Learn; Leave; Lecture; Let; Let on; Let down; Let in; Let in on; Let off; Let pass; Let slide; Let up on; Libel; Lie about; Lie low; Light into; Like; Limit; Link to; Lionize; Listen; Live down; Live up to; Live with; Live without; Look after; Look down on; Look for; Look out for; Look over; Look to; Look up to; Look up with; Lord over; Lose; Louse up; Love; Lower; Lure; Lynch;

Madden; Maintain; Make; Make fun of; Make up to; Make light of; Make up with; Malign; Malinger; Manage; Manipulate; Marginalize; Maroon; Match; Meddle; Mediate; Meet; Mellow out; Menace; Mention; Mentor; Mess around; Mess with; Mess up; Mete out; Mimic; Mind; Mingle; Minimize; Mirror; Misbehave; Miscalculate; Miscommunicate; Misconstrue; Misinterpret; Misjudge; Misread; Miss; Misstate; Misstep; Mistake; Mistreat; Mistrust; Misunderstand; Misuse; Mix up; Moan; Mock; Moderate; Modify; Model; Mould; Molest; Monger; Monitor; Mope; Moralize; Mortify; Mother; Motivate; Move; Mutter; Muzzle;

Nag; Nail; Name; Narc on; Natter; Need; Needle; Neglect; Negotiate; Nickname; Nitpick; Nix; Nod; Notice; Nudge;

Obey; Object; Obligate; Oblige; Observe; Obsess; Offend; Ogle; One up; Opine; Oppose; Opt out; Order; Organize; Ostracize; Oust; Out; Outlaw; Outmanoeuvre; Outwit; Overcome; Overestimate; Overlook; Overpower; Overreact; Oversee; Overstep; Overwhelm;

Pacify; Pain; Pan; Panic; Pardon; Parody; Parent; Parrot; Part; Party with; Pass on; Pat; Patch up; Patronize; Pay; Pay back; Pay for; Peeve; Peg as; Penalize; Perceive; Perform; Permit; Perplex; Persecute; Persist; Persuade; Perturb; Pester; Petrify; Pick; Pick at; Pick on; Pigeonhole; Pillory; Pin down; Pinch; Pipe down; Pipe up; Pique; Pity; Placate; Plague; Play on; Play up to; Play with; Plead; Please; Pledge; Point at; Point out; Poke; Poke fun at; Polish; Pooh-pooh; Poormouth; Posture; Pout; Praise; Pray for; Preach at; Predict; Preen; Prefer; Prejudge; Prejudice; Preserve; Press; Pressure; Presume; Pretend; Prevail upon; Prevent; Primp; Prize; Prod; Profane; Profile; Prohibit; Promise; Promote; Prompt; Protect; Protest; Provoke; Pry; Psych out; Puff up; Pull for; Pull out; Punch; Punish; Pursue; Push; Push around; Put behind; Put down; Put on; Put out; Put right; Put up with; Puzzle;

Quarrel; Question; Quibble; Quiet; Quit;

Rag; Rage; Rail against; Raise; Rank; Rankle; Rant; Rat out; Rate; Rationalize; Rattle; Razz; React; Read; Readjust; Reason; Reason with; Reassure; Rebel; Rebuke; Receive; Reciprocate; Recognize; Recommend; Reconcile; Recruit; Rectify; Redeem; Refine; Reform; Refrain; Refuse; Regard; Regret; Rehabilitate; Rein in; Reinforce; Reinstate; Reject; Relate to; Relent; Relieve; Rely; Remember; Remind; Reminisce; Renege; Renounce; Repay; Repel; Repent;

Reply; Report; Repress; Reprimand; Reproach; Reprove; Repudiate; Repulse; Request; Require; Rescue; Resent; Reserve; Resist; Resolve; Respect; Respond; Restrain; Restrict; Retaliate; Retort; Retrain; Reveal; Revenge; Revere; Revile; Reward; Rib; Ride; Ridicule; Right; Risk; Rip into; Rise above; Roar; Roast; Root for; Rouse; Rub (it) in; Rubberneck; Rue; Ruffle; Ruin; Rule; Run away; Rush;

Sacrifice; Sadden; Salute; Sanction; Sanitize; Sass; Satirize; Satisfy; Save; Scandalize; Scare; Schedule; Scheme; Schmooze; Scoff; Scold; Scorn; Scowl; Screen; Scream at; Screw up; Search for; Seduce; Seek out; Seem; Segregate; Select; Separate; Serenade; Sermonize; Set apart; Set right; Set straight; Settle for; Settle down; Shake; Shake up; Shame; Shape; Shape up; Share; Shirk; Shiver; Shock; Shoot; Shout at; Shout down; Show; Show off; Show up; Shrug; Shudder; Shun; Shush; Shut out; Shut up; Sicken; Side with; Side-step; Sidetrack; Sigh; Silence; Simper; Sin; Sing about; Sing for; Sing to; Single out; Sink to; Sit with; Size up; Slam; Slander; Slant (truth); Slap; Slight; Slip; Slip up; Slow down; Slug; Slump; Smear; Smell; Smirk; Snap out of; Snare; Snarl; Sneer at; Sniff; Snipe; Snitch; Snivel; Snooker; Snoop; Snort; Snub; Socialize; Soothe; Spank; Speak against; Speak for; Speak of; Speak to; Speak up; Speed up; Spellbind; Spit on; Spite; Split up; Spoil; Spoof; Spook; Spout; Spur; Spurn; Spy on; Squabble; Squeal; Squirm; Stab; Stand behind; Stand by; Stand out; Stand up for; Stand up to; Stand with; Stare at; Start; Start without; Stay with; Steam-roll; Step on; Stick up for; Stifle; Stigmatize; Stimulate; Sting; Stink; Stir up; Stomach; Stop; Straighten out; Straighten up; Strain; Stray; Stress out; Strike; Strike back; Strike out; Strive; Stroke; Struggle; Strut; Stymie; Subdue; Subjugate; Submit to; Succeed; Suck up to; Suffer; Suggest; Suit; Sulk; Sully; Support; Suppose; Suppress; Surmise; Surprise; Surrender; Suspect; Swagger; Sway; Swear; Swear at; Swear off; Sweat; Sweet-talk; Sympathize with;

Taboo; Take; Take back; Take on; Talk about; Talk back; Talk into; Talk out of; Talk over; Talk to; Tame; Tamper with; Tap; Tattle; Taunt; Teach; Tear up; Tease; Tell about; Tell off; Tell on; Tempt; Terrify; Terrorize; Thank; Think about; Think of; Think over; Think through; Threaten; Throw; Throw out; Thwart; Tip; Toady; Toast; Tolerate; Torment; Torture; Touch; Train; Trample on; Transform; Transgress; Trap; Trash; Traumatize; Treat; Tremble; Trick; Trouble; Trounce; Trust; Try; Tsk at; Turn; Turn away; Turn down; Turn in; Turn on; Tweak; Twit; Tyrannize;

Uncover; Underestimate; Undermine; Underrate; Understand; Undervalue; Unite with; Unsettle; Upbraid; Uphold; Uplift; Upraise; Upset; Urge; Use;

Validate; Value; Venerate; Veto; Vex; View; Vilify; Vindicate; Violate; Vote for; Vouch for; Vow;

Waffle; Wager; Wait for; Wait up; Walk away from; Walk over; Walk with; Want; Warn; Waste; Watch; Wave at; Weep; Welcome; Wheedle; Whimper; Whine; Whip; Whip up; Whisper; Whitewash; Wimp out; Win; Wince; Wink at; Wise up; Wish; Withdraw; Withhold; Withstand; Wonder; Work on; Work out; Work with; Worm in; Worry; Wound; Wow; Wrangle; Write about; Wrong;

Yell; Yield;

Zap.

Chapter 3

Remember How to Organize: Heterarchy Across Disciplines

Carole L. Crumley
University of North Carolina

Complexity theory is the study of dynamic nonlinear systems, that is, systems that are not in equilibrium and do not act in a predictable manner. Its beginnings in the 1940s are closely entwined with World War II. Complexity theory was developed to address significant problems in the fields of cryptography, cybernetics, and computer design during and after the war. Today complexity forms a coherent subject, with applications in every field of study, finding particular utility when its principles are applied to the biophysical, dissipative system that is the planet Earth.

It is now widely recognized that human societies contribute significantly to earth system dynamics (Table 3.1; Çambel 1993; Steffen, et al. 2004). Fortunately for the social sciences, complexity theory is rooted in fields broadly concerned with models of language and communication rather than earth system science, as applications to human activity are not particularly difficult. For example, key universal features of complexity theory are **integration** (holism, the idea that a system cannot be represented by a simple description of its parts but exhibits emergent behaviour), **communication** (the sharing of information among elements) and the determinative power of **history/initial conditions** (termed chaos or surprise). These features correspond with key features of social systems: the holistic nature of culture (integration), knowledge sharing through the senses such as language, writing, and education (communication), and the formative power of traditions, structures and materials, strategies, and habits of mind (history/initial conditions).

As complexity theory and its attendant vocabulary have become familiar, new applications have appeared in every social science discipline. Many such applications are quantitative and employ nonlinear dynamics, chaos theory, fractals (Byrne 1998; Kiel and Elliott 1996, 1999), and agent based modelling (Gimblett 2002; Gumerman and Gell-Mann 1994; Gumerman and Kohler 1994; Kohler and Gumerman 2000; North and Macal 2005; Soltis, et al. 1995). But complexity theory also lends itself to qualitative modelling, and may also be used at all levels of abstraction – mechanisms and metaphors – including historical narratives (Table 3.2; Bloom 2000; De Landa 2000; Gunderson and Holling 2002; Harvey and Reed 1996; Hornstein 2005; Jantsch 1982; Kauffman 1993, 1995; Mithen 1996; Schieve

Table 3.1 Characteristics of complexity

Complexity and complex systems
- Natural and man-made, as well as social
- Very large, very small, or having both components
- Regular or irregular in physical form
- More likely to occur when there are many parts
- Energy-conserving or energy-dissipating
- Both deterministic and random characteristics
- Aproportional causes and effects
- Linked and synergistic parts
- Positive and negative feedback
- Scales of complexity are relative
- Can exchange material, energy, and information with their surroundings
- Tend to undergo irreversible processes
- Dynamic and not in equilibrium
- Not well behaved, with frequent sudden changes
- Paradoxical
    - Fast and slow events
    - Regular and irregular forms
    - Organic and inorganic bodies in cohabitation

Source: A. B. Çambel, *Applied Chaos Theory: A Paradigm for Complexity* (1993:3-4).

and Allen 1982; Scott 1991; Wheatley 1994; Wilson 1998). Highfield remarks that 'complexity is a watchword for a new way of thinking about the *collective* behaviour of many basic but interacting units' (1996:7).

While several areas of complexity theory (such as chaos theory and basins of attraction) have enormous potential for exploration, in this chapter I concentrate on applications of heterarchy, which treats the diversity of relationships among system elements. While the term may be relatively new to contemporary scholars, heterarchy is an idea that has deep antiquity in human societies and is therefore of special interest to archaeologists and historians. I briefly review pioneering work in heterarchy, then explore its use in archaeology and in other disciplines. I conclude with two examples of how heterarchy can reconfigure thinking about power relations in historic and contemporary societies.

Positioning Heterarchy in Social Theory

Definitions of heterarchy are remarkably consistent across a variety of disciplines, but the work they do is extraordinarily diverse. The earliest definition is from brain research, where McCulloch (1945; see Figure 3.1) contrasts a hierarchy of (ranked)

Table 3.2 Paradigm showing general range of fit between modelling strategies and different levels of ontological complexity in the social system

Hierarchy of Ontological Complexity in Social System	**Modelling Strategies for Studying Chaotic Social Systems: Arrayed by Decreasing Determinist Presuppositions (horizontal axis) and by Levels of System Specificity (vertical axis)**					
Process II: Societal evolution via historical modes of production						
Process I: Class struggle. Conflict over cultural hegemony						
12. Values II: Hegemonic culture & subcultural bases of resistance						
11. Values I: Struggle of hegemonic vs. subterranean world views						
10. Norms II: Allocation of relative power among social institutions						
9. Norms I: Personal conformity to general hegemonic standards						
8. Roles II: Intraorganization allocation of roles & resources						
	Predictive Modelling	Statistical Modelling	Iconological Modelling	Structural Modelling	Ideal Type Modelling	Historical Narratives
	Levels of Modelling Abstraction					

Table 3.2 (continued)

Hierarchy of Ontological Complexity in Social System	**Modelling Strategies for Studying Chaotic Social Systems: Arrayed by Decreasing Determinist Presuppositions (horizontal axis) and by Levels of System Specificity (vertical axis)**					
7. Roles I: Distribution of material rewards & esteem						
6. Facilities II: Technical division of labour in productive sphere						
5. Facilities I: Sociotechnical infrastructure of organization						
4. Ecological organization of institutional time & space						
3. Ecological organization of local biotic community						
2. Biological evolution as a series of assisted bifurcations						
1. Determinant regularities of the physical universe						
	Predictive Modelling	Statistical Modelling	Iconological Modelling	Structural Modelling	Ideal Type Modelling	Historical Narratives
	Levels of Modelling Abstraction					

Source: Adapted from D. Harvey and M. Reed, *Social Science as the Study of Complex Systems* (1996:307 Fig. 13.1).

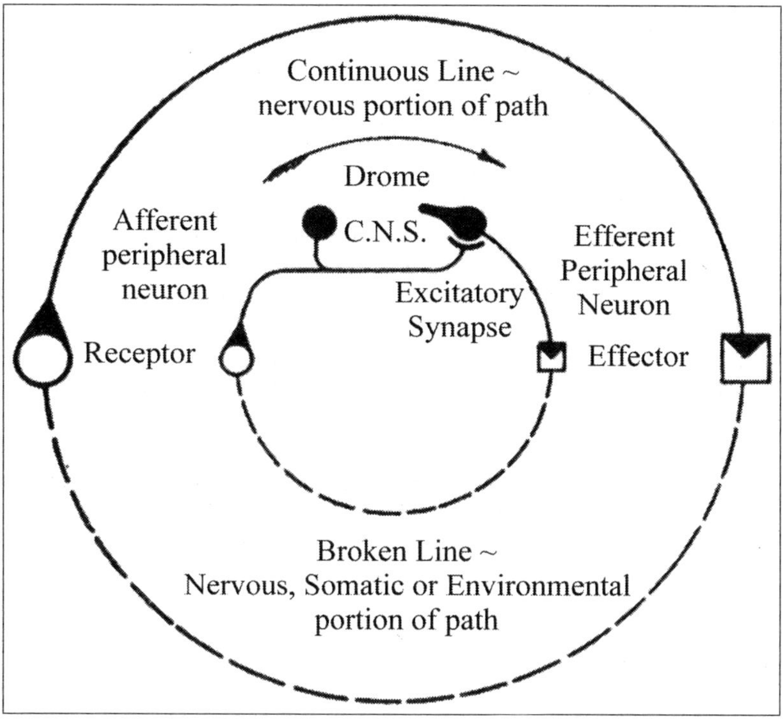

Figure 3.1 A heterarchy of nervous nets

Source: Reprinted from *Bulletin of Mathematical Biophysics*, vol. 7, W. S. McCulloch, 'A heterarchy of values determined by the topology of nervous nets', p. 90, 1945, with permission from Elsevier.

values with a heterarchy of values that defies both ranking and predictability. In artificial intelligence and computer design, Minsky and Papert (1972) refer to the organization of computer sub-routines that can call one another as heterarchical. Mathematician Hofstadter (1979:134) follows Minsky and Papert and defines heterarchy as a program in which there is no 'highest level'. Sociologist David Stark, who studies enterprise in the former Soviet bloc, defines heterarchy as 'an emergent organizational form with distinctive network properties...and multiple organizing principles' (Stark 2001). Social theorist Kyriakos Kontopoulos, following Wimsatt (1976), defines heterarchy as 'a partially ordered level structure implicating a rampant interactional complexity' (1993:381). In archaeology, I have offered a general purpose definition: the relation of elements to one another when they are unranked, or when they possess the potential for being ranked in a number of different ways, depending on systemic requirements (Crumley 1979:144, 1987:158). Heterarchy is both a structure and a condition. Together these definitions call attention to the potential of the system for organizational diversity and change. In general, heterarchical relationships are implicated in the dynamic effect of *difference*, be it spatial, temporal, or cognitive.

Heterarchy does not stand alone but is in a dialectical relationship with hierarchy (where elements are ranked), although from a mathematical standpoint heterarchy is the more general category and subsumes hierarchy as a special case. While heterarchy may be modelled mathematically, it need not rely on mathematical or spatial representations for its application and can be utilized as an abstract model or in historical narrative. In exhibiting this flexibility it resembles the overarching complexity concept of emergence or self-organization, which also supports a wide range of modelling strategies.

Heterarchy meets three criteria upon which any social model must be judged (Byrne 1998:46; Mouzelis 1995). They are: (1) How adequate is the model in relating the micro (individual) level to the macro (social) level? (2) How adequate is the model in relating the conscious agency of social actors to the social structure in which they operate? (3) Can it provide an explanation for discontinuous and fundamental changes in the social system as a whole? Heterarchy can be said to meet the requirements for a robust social theory (Kontopoulos 1993; Mouzelis 1995).

Complexity in Archaeology

While both the new complex systems thinking and the systems theory of the mid-twentieth century (roughly the 1930s through the 1970s) address the organization of information, the contrast between them should be noted. The earlier paradigm was a cornerstone of the New Archaeology during the 1960s and 1970s, offering the tantalizing possibility to many archaeologists that a predictive science of human behaviour could be framed in the language of mathematics and philosophy (Binford and Binford 1968; Flannery 1972; Watson, et al. 1971). Fuelling this interest in North America was the commercial demand for archaeologists after the passage of federal and provincial legislation concerning history, archaeology, and the environment (Ferris 1998; Patterson 1995:108). There were parallel trends in ecology (for an overview see Ellen 1982) and elsewhere in the biological sciences.

Archaeologists who were more interested in cultural historical, cultural materialist, and critical theory topics (especially class and gender) were from the outset uncomfortable with the use of rather mechanistic models to study the human past. Their criticisms eventually drew attention to systems theory's unilinear, positivist, functionalist, and determinist assumptions, along with the practical inability of the models to address system history, change over time, and individual and group diversity (Stein 1998:4).

While there is still some work to do in explaining precisely how complexity theory can address the old objections to systems theory, it is already evident that the new approach readily accommodates history, human cognition, and agency. Heterarchy, while not the only useful concept in the new lexicon, is particularly suited as a corrective to the characterization of power relations in systems theory, which conflates hierarchy with order (Crumley 1987, 1995, 2001, 2003, n.d.a., n.d.b.; Crumley and Marquardt 1987). Archaeologists, already familiar with 'old' systems thinking and its critique, can find refreshing potential in dynamic complex systems research, which offers a means by which human history and individual agency can be accommodated in a non-reductionist framework.

Developing a Critical Approach to Social Organization

Since archaeology's founding as a discipline, several vocabularies for discussing the organizational characteristics of society have been employed, although their implicit assumptions were rarely explored. Those which most consistently dominated interpretation assumed a linear progression from small, early 'simple' societies to those that were more populous, later in time, and 'complex'. Nineteenth century applications of this scheme offered scientific 'proof' that hegemony was the reward for social progress and that the lot of backward indigenous populations around the world could be improved by the colonial enterprise.

Twentieth century American archaeology, in embracing ethnologist Elman Service's (1962) framework of band, tribe, chiefdom and state, eschewed the worst of the nineteenth century assumptions but demonstrated little reflexivity in continuing to use the simple/complex distinction (Patterson 1995). This was especially true in the study of political systems, which were assumed to be more stable the more they tended toward tiered hierarchies of power.

For many archaeologists and university departments in North America the search for and excavation of complex chiefdoms and states became the 'gold standard'. For them, status in the discipline was almost mystically tied to the vanished elites of large polities. Despite considerable ethnographic evidence to the contrary, the archaeological interpretation of state hierarchies as the culmination of ordered progress went unquestioned for several decades. Because the ethnographic record holds organizational schemes of great variety and complexity, this contributed to the mutual scorn between North American archaeologists and sociocultural anthropologists. Archaeology in Great Britain, influenced by geography and history more than anthropology, and with rather different uses for ethnography, bifurcated into Processualist and Postmodernist camps. Neither of these latter groups was quite as obsessed as were North American archaeologists with the epistemology of chiefdoms and states.

Service himself, an ethnographer and ethnologist, drew attention to the importance of coalitions, federations, leagues, unions, and communities in societies of all sizes. Other ethnographers reported immensely complex kin and exchange networks (e.g., Elkin 1964). North American archaeologists, who defined states as elaborate political hierarchies, concentrated instead on how hegemonic power pyramids are constructed by elites. This is understandable in a practical sense, as large sites with monumental architecture are easier to find, although they do not yield so easily to interpretation. Yet as the September 11 events demonstrate, power flows in many channels (Samford 2000) and can manifest entirely outside the framework of state hierarchies and beyond their control. In self-organization terminology, this is termed *chaos* or surprise (Crumley 2001, 2003, n.d.b.), and is related to a characteristic of very hierarchical societies: systemic negligence in intelligence gathering and in recognizing and engaging other dimensions of power (Table 3.3).

In the last decade, the dialectical relationship between the linked concepts of hierarchy and heterarchy has been explored by archaeologists in many regions of the world (Crumley and Marquardt 1987; Ehrenreich, et al. 1995; McGuire and Saitta 1996, 1998; McIntosh and McIntosh 1999; McIntosh, et al. 2000; Rautman 1998; Schoenfelder 2003; Silverman 2002; Stein 1998; Yoffee 2005). While dissatisfaction with the Service model has been a theoretical reason for exploring other models, the greater incentive has been the model's poor fit with much archaeological evidence.

Complex thought and behaviour may be found in every society, past and present. The archaeological record indicates that from earliest human societies to the present day, coupled individual creativity and collective flexibility have met with success. Thus biological diversity has a correlate in human societies: the toleration of difference in individuals and groups increases societal choice and offers a reserve of alternative knowledge for use in problem solving, just as genetic and biological diversity increase ecosystemic resilience. Similarly, organizational flexibility – economic, social and political – enables societies and organizations to adjust to changed circumstances. If we begin with the premise that the tension between competition and cooperation exists in all human societies, it then behooves us to explore the ways rules and norms preserve or deny each, and how both interact with history and changing conditions to forge institutions (Chapman 2003). It is this 'clean(er) slate' that makes heterarchy attractive to researchers.

Social Complexity

From a heterarchical perspective, sources of power are counterpoised and linked to values, which are fluid and respond to changing situations. This definition of heterarchy and its application to social systems is congruent with how the brain works, and was first employed in a contemporary context in the examination of independent cognitive structures in the brain, whose collective organization McCulloch describes as heterarchical (1945, 1989). He demonstrates that the human brain is not organized hierarchically, but adjusts to the re-ranking of values as circumstances change. McCulloch's heterarchical 'nervous nets', source of the brain's flexibility, is a fractal (same structure at a different scale) of the adaptability of fluidly organized, highly communicative groups.

For example, an individual may highly value human life in general, be *against* abortion rights, but be *for* the death penalty (or vice versa). SUV owners may greatly value 'nature' while endangering it. The context of the inquiry and

Table 3.3 Characteristics of authority structures: Hierarchies and heterarchies

Authority Structures

1. Advantages of Hierarchy
   - Clear decision making chain
     · Respond well to fast-developing crises
   - Rules and responsibilities known to all
     · Political interactions few and formalized
     · Political maintenance of the system is low
   - Powerful means of security
     · Defend the organization
     · Suppress internal dissent

2. Disadvantages of Hierarchy
   - Slow movement of information to the top
     · Especially true of subversive activity
     · Formal and elaborate internal security
   - Expedient decisions not necessarily popular
     · High popular dissatisfaction
     · Considerable investment in coercion
   - High security costs

3. Advantages of Heterarchy
   - Good quality information
   - Fair decisions reflect popular consensus
   - Variety of solutions to problems presented
   - Contributions of disparate segments valued
     · Women, ethnic groups, etc.
     · Better integrated group
     · Proud and energized workforce

4. Disadvantages of Heterarchy
   - Consensus is slow
   - Dialogue requires constant maintenance
   - Cacophonous voices and choices

Table 3.3 (continued)

Authority Structures (cont.)

5. Tradeoffs
   - Heterarchical organizations
     - · Value spontaneity
     - · Achieved status builds individuality
     - · Define power as inclusive or counterpoised
     - · Value flexibility and group involvement
     - · Greater response choice/Slower response time
     - · Long-range planning more difficult
   - Hierarchical organizations
     - · Value rule-based authority
     - · Social distinctions elaborated
     - · Power defined as control
     - · Value exclusivity and the *status quo*
     - · Heavy cost for security

6. Democratic organizations
   - Characteristics of both hierarchy and heterarchy
   - More stable than authoritarian organizations

Source: C. L. Crumley, *Communication, Holism, and the Evolution of Sociopolitical Complexity* (2001).

changing (and frequently conflicting) values (Bailey 1971; Cancian 1965, 1976) mitigate this logical inconsistency and is related to what Bateson (1972) terms a 'double bind'. Priorities are re-ranked relative to conditions and can result in major structural adjustment (Crumley and Marquardt 1987:615-617).

Hierarchies and heterarchies of power coexist in all human societies, including states. Societies in which heterarchical values and institutions are dominant are richly networked structures where multiple scales and dimensions are in communication with one another, a condition De Landa (2000) calls a 'meshwork'. The power of various factions and individuals fluctuates relative to conditions; one of the most important is the degree of systemic communication. Societal dilemmas in which values are in conflict are resolved by achieving a novel, transcendent condition in which competing values are re-ranked at particular scales of time and space. At each successive level of integration and over time, new ordering principles come into play. Thus, conflict or inutility leads to *suspension of old forms* but ensure the *preservation* of useful elements to provide creative new solutions to challenges (*transcendence of older forms*). It is in these novel forms that societies retain near-term flexibility, although there is of course no guarantee that the new form is more stable than the old or that tensions will not re-appear in another guise (surprise). For example, revitalization movements such as the Ghost Dance, early Christianity or the 'born again' phenomenon in fundamentalist Christianity seek transcendence through individual and collective rededication based both on new information and the retention of selected old values; these responses require a re-ranking or replacement of values, or mazeway reformulation (Wallace 1970). In sum, the concept of heterarchy indicates an approach to identifying ranked and unranked values, behaviours, and organizations as they shift in time, space, and cognitive frame. As in ecology, researchers must remain aware of intensity, periodicity, and duration of relations; in human societies this might be thought of as the range of powers an individual or group has, and the regularity and duration of service in the managerial role.

Heterarchical Models at Multiple Scales

I now turn to uses outside archaeology which enrich and extend the use of heterarchy. They are drawn from clinical and social psychology, sociology, and anthropology, and from leadership studies in management.

In a groundbreaking dissertation, Lo (2005) has explored heterarchy at several key scales: interpsychic processes and relationships between individuals and within communities, corporations, and polities. She treats heterarchy as a flexible meta-model of order and demonstrates how the model's application could shift power relations if an appropriate means of revealing it (which she terms a 'fulcrum') can be identified. Drawing on McCulloch's brain research on values, she argues that both the construction of self and the patient-therapist dyad could be strengthened by the application of heterarchical principles that are power-neutral or power-sharing. In clinical settings, for example, therapist and patient could work together toward patient well-being (in a relationship rather like that of lawyer-client) rather than the therapist exhibiting controlling behaviours (therapist sits, patient lies down).

Lo's work focuses on taken-for-granted practices of domination and submission and of agency and passivity. In communities, for example, she argues that home hospice help for terminally ill patients retrieves power from hospitals where care is impersonal, as well as restoring dignity for patients and their relatives. In the workplace, Lo examines the underlying system of inequality and how her 'fulcrum' could be applied to improve both productivity and workplace satisfaction.

Research in institutional, organizational, and corporate settings has particular value for archaeologists. Gronn (2004) examines the history of leadership studies in the social sciences, noting that the 'focused leadership' model, although seriously flawed, has long dominated the field (Rost 1993). Other approaches were tried in the 1970s, but for the most part researchers consistently chose the solo or stand-alone leader as their unit of analysis, elaborating on the leader-follower model but failing to examine other forms of leadership. By the 1990s, there was much dissatisfaction with the dualistic nature of the model, and concern that a focus on charisma returns to old 'heroic' models. 'Anti-leadership' researchers argued that other factors were more important (group values, the nature of the work) and refocused research on the investigation of distributed decision-making, distributed cognition, and shared or dispersed leadership. Gronn quotes Yukl (1999:292-293), a critic of the 'heroic leader' paradigm, who asserts that distributed leadership

> does not require an individual who can perform all of the essential leadership functions, or a set of people who can collectively perform them. Some leadership functions (e.g., making important decisions) may be shared by several members of a group, some leadership functions may be allocated to individual members, and a particular leadership function may be performed by different people at different times. The leadership actions of any individual leader are much less important than the collective leadership provided by members of the organization.

This quote should remind archaeologists of Brumfiel's concern about the plethora of uses for heterarchical thinking (1995:125). As Rautman notes, the concept of heterarchy 'cannot be thought of as yet another category or endpoint on a continuum' (such as egalitarian/hierarchical) but 'forces us to specify more clearly the context and temporal duration of the relationships we are describing' (1998:328). Rautman urges researchers not to be discouraged by the variety of applications, but to seize upon the rich fabric of time, space, and domains as a practical guide to research.

Heterarchical Organizing: The State and Beyond

Distinction has consequences. 'A' presumes 'not A', 'self' presumes 'other' and hierarchical organization presumes resistance. In populous societies, this dynamic tension is expressed through the struggle between values that promote consolidation of power in elite hands and democratic, egalitarian, community-based values, which may be expressed in many forms (Scott 1987).

As Scott (1998) has demonstrated in his analysis of 'high modernist' goals of the state, bureaucracies have little appreciation for local knowledge and consistently undermine individuals' capacities for self-governance. The administrative perspective is to see formal order as a precondition of efficiency, hence *visible* regularities are highly valued; nonetheless the population can make its resistance felt through less obvious means: through individual collaboration (increased networking) and functioning, representative institutions.

On the basis of his extensive research, Gerlach (2001) argues that social movements are integrated through heterarchic social linkages among participants as well as the understandings, identities and opponents the participants share. The characteristics of such organizations are that they are segmentary (composed of changing, diverse groups), polycentric (multiple, competing leaders and centers of influence), and networked (integrated network with multiple linkages) and has coined the acronym SPIN (2001:289).

What are such organizations, then, but a democratic challenge to hegemony? State control necessitates the simplification and schematization of information so that it is applicable in many places, but in doing so the quality of information is compromised, and relevant (even critical) distinctions are ignored (Scott 1998:81). This may well have been an important factor in the collapse of Classic Maya power. While the Maya *political* system was organized vertically, the *economy* appears to have been shaped by environmental constraints and characterized by fluidly networked interregional exchange. As water resources, forests, and soil fertility diminished, corporate groups creatively managed food production. Despite the success of such community structures in the countryside, the huge centres of population were apparently ignorant of the corporate role in the conservation of environmental resources. Ultimately, an uninformed attempt at hierarchical management of resources, combined with the insensitivity of urban elites to the fragility of the environment and to the importance of the rural corporate infrastructure, may have crashed the system (Crumley 2003; Scarborough, et al. 2003).

Ancient and Ever-New Forms of Organization

With or without power elites, the organizational principle that is glossed by the inclusive term heterarchy is a fundamental characteristic of complex systems. From an evolutionary standpoint, the first life forms exhibit collective *emergent* behaviour, and even dominance hierarchies among our primate relatives are subject to subversion (Johnson 2001). Although it is possible that the archaeology of small human groups has simply not yielded evidence of vertical power relations, most research suggests that cooperative behaviour was more important to group survival and that the collectivity necessarily suppressed dreams of individual power. In regard to states, Stein remarks that

> The dynamics of conflict between the centralized elites and other social sectors not only define the structure of the polity, but also help explain when and why evolutionary change takes place. To understand these processes, we need to explore the dynamic, fluid nature of power relationships and their longer-term transformations... [by focusing on] conditions under which power relationships... undergo major structural transformations. (1998:26)

Archaeologists can profit from contemporary studies of both constrained situations (such as the workplace) and broader arenas where disaffected agents are freer to creatively organize. Workplace and organization-based studies reveal the deep hold hierarchy has on contemporary imagination, but they also point the way toward educational reform (Clarke 1972:5ff.; Gronn 2004; Lo 2005:151ff.).

Organizational activities that reach beyond the framework of the office, corporation, or nation-state are interesting for other reasons. After the events of September 11, 2001 in the United States, the shadowy organization of Al-Qaeda became front-page news. Neither centralized and bureaucratic nor amorphous, this excellent example of Gerlach's SPIN organizations caught the American defence community by surprise, in large part because the intelligence sector knew little about such organizations or the Middle East. The US mounted a traditional Cold War response and attacked nation-states (Afghanistan, Iraq), sought 'leaders' (Osama Bin Laden, Saddam Hussein), belatedly addressed intelligence issues (9/11 Commission Report 2004), and has yet to recognize in Al-Qaeda a new kind of adversary (Arquilla and Ronfeldt 2001; Griffin 2004). Of central importance for the success of Al-Qaeda has been the Internet, freeing its activists from temporal and spatial constraints and intensifying the exchange of information. This is also true of the Zapatista 'social netwar', the world's first 'postmodern' movement (Arquilla and Ronfeldt 2001:189).

Although I agree in large measure with Ronfeldt (1996) who, like Jantsch (1982) sees this trend in evolutionary terms (Figure 3.2), I would argue that highly networked SPIN organizations have always been able to elude hierarchical institutions and operate beyond and within their boundaries. Excellent examples are the revitalization and millenarian movements mentioned above, as well as every major religion. Somehow the values that underlie them have been transmitted, even without the Internet.

One of the old forms has been anarchy. Far from meaning *no order*, it specifies an order and set of values that are explicitly local, consensual, without leaders/followers, and *not* hierarchical. Bose (1967) traces anarchist thought back thousands of years to the first kingdoms and states, finding anarchist ideas in the work of the Chinese philosopher Laotse, Greek cynics, and Roman stoics. Were one to add in the numerous utopian thinkers, the list would be long.

While scarcely characterizing all resistance movements everywhere (because many who resist wish to seize power rather than neutralize it), anarchism does offer a consistent and coherent philosophical argument against the various forms of organizational hierarchy. It is hardly surprising, then, to find that globalization has revitalized anarchist thought while chaos theory and the Internet have facilitated anarchist practice (Table 3.4; CrimethInc. Workers' Collective 2001; Graeber 2004; Ludlow 2001; Post 2001; Purkis and Bowen 2004:14).

Conclusions

Because it is a fundamental organizational principle of complex systems, I argue that there is no need to confine the heterarchy concept to a narrow range of applications. At any stage of its application, debate is healthy. Eventually, disparate researchers will become aware of one another and applications of heterarchy will become more consistent and 'basins of attraction' will emerge.

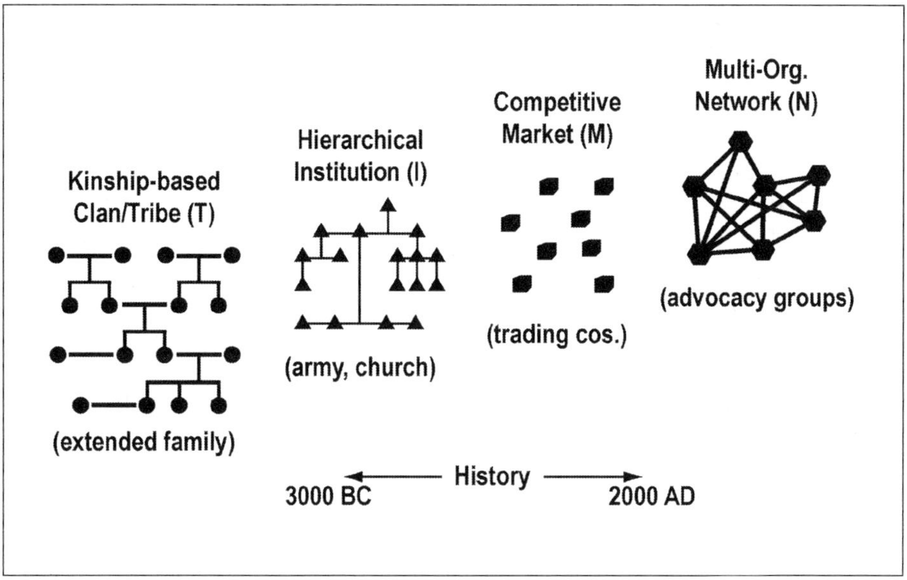

Figure 3.2 Basic forms underlying the organization of all societies

Source: After Ronfeldt (1996:3 Figure 1).

As Chapman reminds us, archaeology must refute a history of our species that finds dualities everywhere – simple/complex, equal/unequal, civilized/uncivilized – and naturalizes inequality (2003). Instead, we must embark on a more difficult task that asks more of both the archaeological record and of our interpretive frameworks.

Do networks leave material evidence? Rhea Rogers (1995) argues that they do, and that we have already finished the hard work of excavating and analyzing, leaving only the need for a fresh interpretation. By treating every artifact category, material, technology, feature and physical location as a separate body of evidence, she found that spatially and culturally delineated 'tribes' of the Carolina Piedmont are better understood as protean, interactive, overlapping networks that were unbounded in terms of both personnel and space.

Does resistance to hierarchies leave material evidence? Perhaps widespread evidence for the collapse of political organizations and the continuity of their populations point to resistors' successes (Diamond 2004; Scarborough, et al. 2003; Tainter 1988). Do social movements leave material evidence? Perhaps that can explain the disjuncture between Middle Woodland societies, widespread in eastern North America, and the more limited distribution of Hopewell symbolism. Must we rely upon the archaeological record alone? Definitely not! We have always had a wealth of ethnohistoric and ethnographic data. Now, thanks to the inclusive framework of complexity theory, we have examples and models from every discipline upon which to draw. Let's get to work.

Table 3.4 Ellickson's Five Controllers that provide substantive rules governing behaviour

Ellickson's Five Controllers

Controller	Substantive Rules	Sanctions
The actor	Personal ethics	Self-sanction
Second-party controllers (i.e., the person acted on)	Contractual provisions	Various self-help mechanisms
Nonhierarchically organized social forces	Social norms	Social sanctions
Hierarchically organized nongovernmental organizations	Organization rules	Organization sanctions
Governments	Law	State enforcement, coercive sanctions

Source: D. G. Post, *Anarchy, State, and the Internet: An Essay on Lawmaking in Cyberspace* (2001:199).

Chapter 4

Agency, Collectivities, and Emergence: Social Theory and Agent Based Simulations

Christopher S. Beekman
University of Colorado at Denver and Health Sciences Center

Introduction

This chapter is intended to situate certain complex systems concepts in relation to agent based theoretical approaches currently employed in archaeology. The chapter, in a sense, builds upon an article by Nigel Gilbert in which he describes the potentially productive relationship between agent based simulation and models of social agency from sociology (Gilbert 1995). In the first part of this chapter, I briefly examine the concept of agency as it has been imported into archaeology, followed by a discussion of sociology's critiques of one of the primary sources from which archaeologists have borrowed agency.

This will draw us into a discussion of emergence as it is understood by both complex systems theorists and other disciplines. Emergence is ultimately the notion that different rules apply at different scales of analysis, and that one cannot predict the rules at one scale from an understanding of the system at a lesser scale. Although emergence is cited frequently in the nonlinear literature and is quite consistent with Ilya Prigogine's seminal work on phase transitions, it is conceptually opposed to another oft cited 'characteristic' of chaos in the literature – that of fractal patterning or self-similarity across scales. This jumbling together of contradictory concepts under the broad umbrella of complexity demonstrates that the field can be viably critiqued as without theoretical consistency. I argue that a more realistic appraisal of agent based simulation as a tool that can represent quite different theoretical schools will allow researchers to begin building bridges that clarify the true usefulness of the approach for archaeology.

With this background, we can suggest changes that would bring current agent based computer simulation more into conceptual agreement with *both* emergence and theories of social agency. These changes include a more explicit approach to hierarchy, collective agency, and the formation of institutions. I suggest that traditional agent based simulations that propose to model all complexity 'from the bottom up' miss the nonlinear effects that occur when we model social institutions with capabilities for downwards causation, and indeed, fail to explain the existence of high level institutions such as the State, which have clear roles in coordinating and dictating action and engage in self-protection (e.g., Lemieux 1995). Most centrally, we must recognize that simulations that purport to empirically model social systems are in fact espousing particular theoretical interpretations of motivations for human behaviour. The longstanding theoretical acknowledgement that data are theory-laden has been slow to receive attention by complex systems theorists.

In the final part of the chapter I make a verbal application of these points to certain problems in the Pre-Columbian archaeology of Jalisco, Mexico. There is growing evidence that partly descent-based social units (such as lineages) held rights to corporate property and were the building blocks of Late Formative-Classic period (200 B.C.-A.D. 500/600) society. While evidence exists for significant social competition between these groups in some contexts, several of these groups formed a higher order social unit with ritual and political responsibilities. By discussing how and why these groups may form out of a field of individuals, and how and why in turn these groups form collectives at still higher scales of analysis, I hope to clarify that 'agents' are far more problematic entities for complex systems theorists than is usually appreciated. The nature of agents changes substantially as one moves further up the scale of organization. A fuller engagement between modellers and social theorists is called for.

Agency in Archaeology

The social sciences have always shared a central interest in the interaction of structure and agency, of culture and personality, of society and the individual. Past approaches from Karl Marx to Herbert Spencer to Emile Durkheim to Claude Levi-Strauss have tended to give priority to structural factors, treating individual humans as relatively reactive, more affected by than producing effects in the social system of which they are a part (Brettell 2002). This top-down approach was most clearly exemplified by Parsonian sociology, by the Functionalist and Structural schools in anthropology, and by the Processualist and Structural Marxist schools in archaeology. Although accompanied by various analytical successes, these approaches have been buffeted by endless critique in recent decades for eschewing human agency. Structural factors have seemed so pivotal for understanding long-term change that many have found it difficult to recognize that people may have been oblivious to those issues in the short term. Individuals often make decisions based on situational, subjective evaluations that may even have been unwise in the medium or long term. Clearly the shorter term, smaller-scale perspective must be better understood, because society is indeed the accumulation of the decisions and actions of local agents, even if their actions are shaped by major structural factors such as the physical environment or social ethos. The Post-Processual contribution to archaeology in the 1980s and 1990s (exemplary texts include Hodder 1991; Shanks and Tilley 1997) was to help shift the scope of archaeological theory towards smaller social groups and their disparate goals, and to make a general break with normative conceptions of behaviour.

It was Ian Hodder (1982) who, actually preceding Ortner's seminal article for cultural anthropology (Ortner 1984), proposed that archaeologists look more closely at existing social theory seeking a better approach to the micro-macro relationship, or the individual in relation to society. This literature on agency includes Practice Theory (Bourdieu 1977, 1992), Transcendental Realism (Bhaskar 1975, 1979, 1989, 1993), Structuration (Giddens 1979, 1984), Morphogenesis (Archer 1982, 1995), Social Becoming (Sztompka 1991), and other approaches (e.g., Granovetter 1973, 1978, 1983, 1985; Sztompka 1994). The various theories deal in some way with the agent, variously described as an individual or small group whose actions instantiate the wider 'rules and resources' that constitute the structure of society (Giddens 1979), but who are in turn constrained or enabled by those structural features in a recursive or dialectical relationship. By buying into (or rejecting) social concepts of rank, race, religion, etc. and acting in concert with those ideas, our actions reproduce (or transform) those concepts, although of course that structure existed prior to us and influenced our decisions. A truism, perhaps, but the theorists cited above differ considerably in how this relationship is to be conceptualized. This shift in focus towards agency has, at the very least, left behind the tired practice of favouring either material or ideological factors as causal in social change. At the scale of individuals, material and idealist motivations are often conflated, and research has moved away from this sterile debate and even from causality in general. There has instead been a burgeoning interest in the distribution of power and the degree of knowledge about the system of which one is a part.

Three basic misconceptions regarding agency must be laid to rest before we may continue. 1) Agency is an outgrowth of Post-Modernism; 2) Agency is about the ability of individuals ('Great Men and Women') to change the world; 3) Studies of agency focus on identifying individual people, almost impossible in the archaeological record. The first misconception stems from the continuing and unhelpful conflation of Post-Modernism with Post-Processualism (e.g., Kuznar 1997). While both schools of thought admittedly resist easy definition, classic Post-Modern scholars such as Derrida or Foucault betrayed little interest or optimism regarding the individual, whereas those social theorists singled out by Post-Processualists as properly considering human agency (e.g., Giddens) consider themselves to be steering a middle course between objectivity and subjectivity. As to the second misconception, structure is made up of the actions of all individuals, conscious or not, and social theories of agency do not even focus on transforming structure. Accommodation is agency too. Finally, even a cursory glance at social theorists who handle agency as a conceptual approach quickly reveals that real individuals receive very little attention (a landmark exception being Archer 2003). Analysis still focuses on aggregate phenomena, but it is *explanation* that must be couched in terms of individual action, since societies as units of analysis cannot 'act'. Agency approaches are most clearly distinguished by the scale at which explanatory narrative takes place. For example, a society cannot adapt to the environment. Rather, individuals make local choices about how they will obtain their food (based on their subscription to social concepts regarding appropriate food, activities to get food, etc.) that result in aggregate patterns at the level of the society, creating the illusion that the society has 'done' something.

Archaeologists following the path laid down by sociologists have tended to distinguish their approach sharply from others that also reflect a renewed interest in the individual. Indeed, in a transparent move towards appropriation, evolutionary ecology, selectionist, and other approaches that have focused on the individual were quickly dismissed from the original defining volume on agency in archaeology (Dobres and Robb 2000:8 – for similar sentiments *vis a vis* rational choice theory in sociology see Archer and Tritter 2000).[1] Of the contributing authors only Clark (2000) felt it necessary to discuss his own approach in relation to two major competing interpretations of individual decision making – Evolutionary Ecology (e.g., Smith and Winterhalder 1992) and Darwinian selectionist theory (e.g., Maschner 1996). The former has established its largely unchallenged dominion over mobile hunter and gatherer studies, while numerous variants of the latter exist. Clark quickly set aside evolutionary ecology from the discussion as too methodologically individualistic, in that individuals are considered to be too rationally calculating and optimizing, and they act without regard for social constraints (i.e., it is a strictly bottom-up approach).[2] He takes more time to conclude that in the end Darwinian approaches tend to emphasize innate human tendencies which become causal in their importance, whereas he describes his favoured approaches from sociology as eschewing causality because of the dialectical relationship between agency and structure.[3] But even the sociological models Clark cites are riven with divisions, especially the two (Archer 1995; Giddens 1979) that he uses to inform his contrast between Darwinian and social theory approaches. I intend to explore those divisions a bit later.

The more commonly cited rift in characterizing human agency is that between the sociological icons Pierre Bourdieu and Anthony Giddens (e.g., Stone 2003), although each position is quite complex and other authors espouse similar approaches. Bourdieu's approach (1977, 1992) incorporates ethnographic data and a rich set of theoretical concepts such as the *habitus* that link the micro and macro scales of analysis. He has tended to depict a rather 'heavy' structure and emphasizes the difficulties in breaking out of the patterns and world views into which we have been acculturated. Bourdieu's actors certainly reproduce structure

---

[1] 'Among other recent approaches exploring individual interests and actions and their contribution to long-term, large-scale social transformations have been optimal foraging theory, varieties of game theory, as well as Darwinian and evolutionary ecology models' (Dobres and Robb 2000:8). This narrow approach has been critiqued by many since (see Gardner 2004a).

[2] An individual with their own vested interests who shows no regard for others or the rules of society – this is the classic definition of a sociopath.

[3] But see Somers (1998) for a detailed discussion of the role of causality in positivism and realism.

through their actions, but largely through the unthinking repetition of daily practices (non-discursive action), and less through conscious reflection and modification of their world. Giddens (1979, 1984), on the other hand, has insisted that actors need to be conceptualized differently, as individuals with vested interests who have some penetration of and act upon structure. Agents do not consciously plan the creation of their social world (a more strictly Social Constructivist argument), but they do act with intent and can explain their motives.[4] Among Giddens' oft-cited statements is 'structure has no existence independent of the knowledge that agents have about what they do in their day-to-day activity' (1984:26). There is a Schrodinger's Cat quality to structure in both Giddens' and Bourdieu's work, in that it is only instantiated through action.

Macy (1998) characterizes Bourdieu's routinized practice as drawing upon the past for behavioural cues, while Giddens' intention based action is more influenced by a consideration of (short-term) future goals. T. Cook (2002:xiii) pithily contrasts the two as 'non-rational commitment or rational instrumentality'. This basic distinction has become more pronounced as archaeologists have increasingly cooked-down the source literature. Explicit formulations of agents as aggrandizing individuals pursuing their own vested interests through goal-seeking action have since become more common (e.g., Beekman 2000; Clark and Blake 1994; Joyce and Winter 1996; Joyce 2000, 2001; Marcus and Flannery 1996). But so have applications of Bourdieu's concept of routinized practice (e.g., Dietler and Herbich 1998; Hodder and Cessford 2004; Lightfoot, et al. 1998; Pauketat 2000). In a sense, the polarization between the two approaches has come to duplicate the very bottom-up methodological individualism (discursive strategizing action) and top-down institutionalism (non-discursive practice) that were supposed to have been brought together under Giddens' concept of Structuration, pointing up the difficulties of finding an appropriate synthesis of society-focused and individual-focused models. Certainly, Archer (1982:458-465) has drawn attention to the fact that Giddens himself portrayed agents as particularly cognizant of structure at some times and unwittingly following its rules at others, without reference to the conditions under which one or the other might predominate. While sociological research has tended towards totalizing narratives, variation across space and time is more heavily emphasized in ethnography and archaeology, and hence we might predict that the importation of these approaches into anthropology would result in problems.

---

[4] Silliman (2001:192) makes this same distinction between characterizations of agency, but he puts both Giddens and Bourdieu together as examples of heavily socialized approaches to agency. To judge from citations, Silliman clearly distances Giddens from those archaeologists who specifically declare their work to be inspired by him. This only underscores the diversity of interpretation possible with Giddens' work and has happened elsewhere in archaeology as well. In a recent volume on archaeological community, one author (Yaeger 2000) presses for the importance of interactionalism as exemplified by Giddens and Bourdieu, without noting that their lack of attention to interaction is one of the major critiques of their work.

A related framework developed in archaeology in recent years is that of a continuum between network and corporate strategies (Blanton, et al. 1996), or individual and group-oriented social strategies. Though clearly inspired by the Giddens/Bourdieu divide, there are differences. The network strategy is one in which motivated aggrandizers compete with one another for power, prestige, and followers through performance and investment in symbolic goods obtained through exchange. The corporate strategy is considerably more faceless and group-oriented, and social roles and membership in broader groups take precedence over individual aggrandizement. In the authors' formulation, each of these strategies tends to be dominant in a society at one moment or another, and they emphasize both cultural and temporal variation in these strategies. Some related work (Beekman 2000) suggests that such strategies can be concurrent but associated with different contexts (e.g., public vs. private spaces). A student (Audrey Al-Ali personal communication 1999) pointed out to me a highly illustrative example from Bali. The Balinese irrigation networks, the group action that sustains them, and the water temples that aid in scheduling and coordinating irrigation through ritual (Lansing 1991) form a corporate or group-oriented strategy in Blanton, et al.'s (1996) terms. Parallel to this phenomenon, however, is the system of interacting and competing political elites and their kingdoms (Geertz 1980). This individualistic, competition-oriented system corresponds to Blanton and colleagues' network strategy, and has no control over the temples or irrigation systems, which form a separate power structure altogether. The Network-Corporate distinction is therefore better thought of as differing sets of rules for social practices in different contexts.

We can take this several steps further. There is a strong gradient in every society between unquestioning actors and those with the training or perspicacity to penetrate structure. But even within the category of individuals, each of us tends to be simultaneously less introspective over some aspects of our lives and highly sceptical and questioning over others. Giddens recognized this and codified these distinctions as practical and discursive consciousness (Giddens 1984:7, 41-45), but archaeological interpretations of Giddens have tended to lose that distinction and cite Giddens instead for those statements he has made regarding agents' intentionality and ability, not their failures to exhibit these qualities. Clearly, selecting one or the other as the theoretical basis of one's approach is distinctly one sided.

A recent article by Gillespie (2001) pointed out that many early studies dealing with agency had overshot the mark. The enthusiasm over the rediscovery of human agency in archaeology had led to its overemphasis. Gillespie noted that Western individualism had played a role here, and that archaeologists had been all too eager to emphasize the solitary actor (constrained or enabled by structure, to be sure) with less reference to the larger social groupings of which the actor was a member (e.g., family, lineage, house, class, faction, ethnic group, polity). Archaeologists who have relied upon Giddens' work in particular have opted for a variant of agency that emphasizes the interpenetration of individual and structure, with little attention to the interaction of agents with one another. Gero comments,

'...the theorized classic liberal agent acts alone and autonomously, unaided (unhindered) by webs of relationships and social networks, discounting intimate relationships, participatory decision-making, or negotiation' (Gero 2000:34).

Gillespie's solution was the concept of personhood – the manner in which an individual is recognized socially by others as part of the wider groupings to which one belongs. This approach still leaves the individual human being the unit of analysis, but as they are seen by others as a reflection of larger social units. The concept is roughly the mirror image of Bourdieu's *habitus*, which is the individual's own view of structure as it has been internalized (embedded structure [Scott 2001:84]) through their personal experience.

Gillespie's work is a turning point in studies of agency. Up to this time archaeologists influenced by agency approaches have accelerated their efforts to focus on the contributions of individuals within the aggregated and sedimented activities of many people (perhaps culminating in Hodder 2000) as a means of understanding the micro-macro problem. Gillespie began with forms of archaeological data (burials) that might plausibly be attributed to individuals, and instead saw a role for collectivities interacting between the extremes of agent and structure. Interaction and social relations are once again worth another look.

Critiques of Giddens' Structuration Theory

The growing schools of Critical Realism and Relational Realism present an approach that differs from Giddens and Bourdieu partly in philosophy and partly in their more pragmatic approach to analysis (e.g., Archer 1982, 1995, 1996[1988], 2000, 2003; Archer, et al. 1998, 2004; Bhaskar 1975, 1979, 1989, 1993; Brown, et al. 2002; Collier 1994; Layder 1985, 1997; Wilmott 1997, 1999). While Giddens represents that thread in sociology that approaches social analysis through semiotics, hermeneutics, and linguistic structuralism (cf. Sewell 1992), the growing Critical Realist school comes from interactionist and systems-based theoretical lineages within the same discipline. Yet Critical Realist disagreements are not those of old debates, nor do practitioners attempt to resurrect exclusively top-down or bottom-up approaches. While subscribing to the recursive relationship between agency and structure, Giddens' opponents find his approach overly philosophical, and critiques of his work frequently note the lack of a useful approach to operationalization.[5]

[5] Stones' (2001) defense of Structuration is not very compelling. He grants Archer's (1982, 1995) methodological improvements while quibbling with the conceptual difference that demanded those improvements. In addressing Structuration's methodological gaps, he finds that some practitioners have developed *ad hoc* analytical mechanisms, missing from Giddens' original formulation, that duplicate those more concretely laid out in Archer's work. I would argue that this is happening in archaeology as well, which begs the question why people continue to cite Giddens as their inspiration. Such citations are honorific.

Margaret Archer (1982, 1995, 1996[1988], 2000, 2003) in particular has laid out a long series of critiques. Many of these criticisms ultimately relate to Giddens' preference for a concept of 'duality of structure', as opposed to the 'analytical dualism' proposed by Archer and others (e.g., Archer 1996[1988]:72-96).[6] For example, Structuration has been critiqued as synchronic (Archer 1982:466-471). Despite Giddens' stated interest in incorporating time and temporal change, his decision to fuse agency and structure into an analytical whole blurs human action and structural transformation into a continuous stream. While certainly justified at some level of thought, Archer complains that such an approach only hinders actual problem-oriented analysis that requires a prior state, social interaction that leads to change, and the new resulting state. Duality also obscures the fact that the structure with which agents must contend is not their creation, but that of those agents who came before them, further teasing these poles apart. Structuration is critiqued as non-propositional, failing to develop an approach that inquires why agents are free to develop new solutions at some times and constrained at others (ibid, 458-465). Giddens' own incorporation of two previously opposed approaches did not resolve them, but merely subsumed both opposing views, and his discussion tended to flip-flop between hyperactive agents and stifling structure (a similar result is found in Sztompka 1991). Again, Archer argues that Giddens has rolled structure and agency into a single recursive duality, and there are no other intervening collectivities with an analytical presence that might allow us to more deliberately dissect the micro-macro linkages.[7]

This leads to a more important issue for our discussion. Giddens' duality of structure means that the theoretical and analytical focus is on the individual's relationship with structure, with very little discussion of the interaction *between individuals*, i.e., actual social relations (Archer 1982:471-477; Kilminster 1991:94-102; Mouzelis 2000; Willmott 1999). Structuration concentrates on how structure becomes embodied in the individual, and constrains or enables individual action that will transform or reproduce structure. A corollary or implication of this point is that Structuration does not recognize the analytical utility of collective agents – temporary and fluctuating groups of individual, interacting agents (lineages, sodalities, factions) with unpredicted *emergent properties* that make these higher order units more than the sum of their parts (Hindess 1986; Mayntz 2004; Sewell 1992:21). Archer (1982:475) notes:

> Emergence is embedded in interaction: in the latter 'we are dealing with a system of interlinked components that can only be defined in terms of the interrelations of each of them in an ongoing developmental process that generates emergent phenomena –

---

[6] The distinction may seem arcane, but has its origins in Descartes and his mind-body dualism. Giddens sees this as a false dichotomy being played out with agency and structure as well. He is primarily concerned that such a dichotomy reifies structure and gives it too much autonomy (cf. Willmott 1999 for the counter-argument).

[7] Andrew Gardner (2004b) is, to my knowledge, the only archaeologist to have noted this line of critique and its relevance for conceptualizing agency at different scales.

> including those we refer to as institutional structure'. Emergent properties are therefore relational: they are not contained in the elements themselves, but could not exist apart from them. As Blau puts it, 'although complex social systems have their foundation in simpler ones, they have their own dynamics with emergent properties'. The latter can arise at all levels from small scale interaction upwards, although as scope grows they are increasingly distanced from everyday psychological dispositions but never ultimately detached from interaction. The highest orders of emergence are nothing more than the relations between the results of interaction. Nevertheless these 'feed back' to condition subsequent interaction at lower levels.

That recursive relationship between agency and structure is therefore repeated at multiple scales where it *emerges* with forms of social organization.

Emergence and Collective Agency

Emergence is a current topic of import to philosophers of mind, sociologists, and complex systems theorists as an alternative approach to the micro-macro link (e.g., Bedau 1997; Cunningham 2001; Germana 2001; Hodgson 2002; Humphreys 1997a, 1997b; Kay 2001; Kim 1992, 1999; Mayntz 2004; Mihata 1997; Newman 1996; O'Connor 1994; Rueger 2000; Sawyer 2001; Schroder 1998; Shoemaker 2002; T. Smith 1997; Smith and Stevens 1996; Welshon 2002; Wimsatt 1997, to name a few). The concept is constantly being redefined using different combinations of basically the same major points. An emergent property is a higher level element in a system composed of many smaller components. Characteristics associated with emergent properties are as follows, although different theorists emphasize different ones; 1) Emergent properties are more than the sum of their parts, 2) they form not from the intrinsic properties of the lower level components, but rather their relational properties, 3) they do not follow the same laws or rules as the lower level components, and 4) they may be capable of downwards causation, affecting the very components that created them.[8] The mind-body problem is often the forum in which the participants couch their debate, but emergence can take place in other kinds of systems and both 'properties' and 'institutions' have been discussed as emergent. Another point of disagreement is whether non-deducibility of emergence is to be characterized as ontological or epistemological, i.e., it may be a real phenomenon or it may just be a reflection of current levels of knowledge.

One interesting implication of studies of emergence is that individual dispositions are themselves emergent in some sense, not from 'above' (from structure), but from 'below'. Consciousness is often given as an example of an

[8] Downwards causation, as a component of recursive social theories like Structuration or Morphogenesis, has been hard for many philosophers of mind to accept (e.g., Kim 1992). It is also far from widely accepted in sociology, as the debates between Social Constructivists, Rational Choice Theorists, Critical Realists, and supporters of Structuration demonstrate. Anthropological theorists have been quicker to accept the idea that different social rules and resources (capable of being equated with culture) impact upon individuals' actions.

emergent property, as it is dependent upon the organization and interaction of biological components which individually obey rules very different from those that describe the mind (e.g., Sperry 1986, 1991). Some neurological and biopsychology studies propose that cognitive activity and the 'self' are similarly supervenient upon the relationship between the biological components of the body and cannot exist without them (Smith and Stevens 1996:141). Indeed, there are suggestions that our higher cognitive functions may have developed through the demands of interaction with other humans (Orbell, et al. 2002). Whether the mind can turn around and affect the body's biological processes is contentious, but succinctly renders what emergentists mean by downwards causation.

The concept of emergence in the social sciences is sometimes associated with interactionists like George Herbert Mead (1934) and Norbert Elias (1969, 1978; see O'Connor and Wong 2002 for the most comprehensive history), but many discuss emergence in relation to the work of Georg Simmel, who saw individuality as emerging through interaction (Wellman and Frank 2001). Simmel saw each of us as having multiple motives at all times, in particular the competing desires for imitation and differentiation (Nedelmann 1990). The desire for social conformity or social innovation is thus bound up within a single person, and Simmel (1971[1908]) saw individual autonomy as a constructed phenomenon connected to one's position within a web of social relations.[9] Appeals to see individual humans as the obvious analytical endpoint of all action have been repeatedly singled out as problematic. Following Simmel's much neglected legacy, Archer (1995:247-293, 2000), Gilbert (1995:149), and Gero (2000:37) all quite rightly make the point that human beings are themselves emergent entities, with internal conflicts and differing sources of influence (see Hardcastle 2000) not captured by Giddens' unconscious, practical consciousness, and discursive consciousness.

In Giddens' foundational theoretical work in the 1970s and 1980s, (1979:50-60) he opposed the notion of emergence because he felt that it reified structure, which had in turn been associated most frequently with constraint in Parsons' sociology. In response to his early critics, Giddens made it clear that structure is both constraining and enabling (1984:162-185). His continued opposition to emergence (e.g., Giddens 1984:171-172) seems to stem from his association of collectivities with functionalism, but it is ultimately inconsistent with Giddens' most basic argument for the recursive relationship between structure and agency. How after all can structure hold up its part of the bargain and have any constraining (or enabling for that matter) effect upon agency unless we grant that the attributes of structure or collective agents take on a reality of their own and become resistant to change through agents' actions? It is possible to argue that denying emergence is tantamount to abandoning the recursivity between agents

[9] With one of my friends I am active and outgoing in contrast to his more reserved nature, but when spending time with another, more boisterous friend I become so reserved that my first friend might not recognize me. Rather that emphasizing my own individuality and 'me-ness', this phenomenon suggests that Simmel's interactionist view is far more useful.

and structure, and returning to a methodological individualism. Willmott (1999:12-13) directly accuses Giddens of this, in fact.

Giddens' notion of structure as rules and resources has been important for understanding the system with which actors must contend. But he viewed institutions as highly transitory. He described them variously as 'deeply layered practices' (1979:65), as 'stretched' social activities (1984:xxi), and as 'articulated ensembles' (1984:170). At one point Giddens recognizes that institutions are qualitatively different and obey different rules than individuals (1984:177-178), but the epiphany fades and his treatment of organizations *in relation to individuals* is commonly brief (e.g., Giddens 1990:302-304). What is partly the problem here is that rules and resources vary between different institutions. The different groups to which one belongs or the institutions within which one attempts to act each set the constraining or enabling conditions for agents, not the more distant and evanescent 'structure'. In other words, structure is differentially actualized through groups or institutions, which in turn have a recursive relationship with their actor-members (Vaughn 1999). It is thus at the scale of social groups, institutions, and collective agents that emergence has the greatest potential for clarifying the relationship between structure and agency.

A reorientation of agency to include emergence argues in favour of a stratified approach in which we attempt to identify different scales of society at which different rules of action predominate (Brante 2001; Spencer-Wood 1990, 1996). Interaction creates groups. Groups in turn can become institutionalized and partially divorced from the very social actors whose actions created them. Institutions may be created through regularized practice, but by definition have become partly decoupled from practice. Thus disarticulated from individuals, an institutionalized group becomes not only more robust, but also subject to different rules and its interests cannot be glossed as the negotiated interests of the aggregate of its members.

For example, an interesting study compared the campaign contributions of corporations to the contributions made by individual capitalists who were associated with those same organizations (Burris 2001). The donation patterns were quite different between the two scales of 'actors'. The 'motives' of corporations could not be explained as a negotiated aggregation of the motives of their constituent members. Corporation contributions were more pragmatic and geared towards purchasing influence with political officials, regardless of party affiliation, whereas the donations of individuals were more variable and reflected a more direct ideological interest in securing the election of a favoured candidate. Some of these differences were due to different social rules (read structure) applicable to organizations as opposed to individuals; but while psychological motives were important for individuals (needs, desires, emotions), the corporations' actions were more clearly understood in emotionless, rationalized terms emphasizing self-preservation. Prestige, an ultimately emotional goal for individuals, was a meaningless category for understanding organizational behaviour.

Emergence and the decoupling of institutions from their member actors results in a much greater emphasis upon discontinuities and nonlinear reconstructions as one works up from individuals through institutions. Archer's own rather complicated formulation begins with the self, the person, the actor, and the agent (Archer 1995:247-293, 2000:253-305, 2003; compare the breakdown of agency in Emirbayer and Mische 1998), not all of which I care to pursue here.[10] Agents are for her collectivities with vested interests of some kind. Primary Agents are those 'members of collectivities who share the same life chances' (Archer 2000:11) and are part of the same class, ethnicity, or other social category. Primary Agents may well have vested interests, but they are unable to alter structure through purposeful action due to the lack of organization, effective power, etc. This lack of coordination among 'unaligned' agents does not vitiate their potential effects upon structure, but it strictly hinders conscious goal-seeking action. 'Yet similarities of response from those similarly placed can generate powerful, though unintended aggregate effects which is what makes everyone an agent' (Archer 1995:259).

Institutions and other organized groups are what Archer dubs Corporate Agents. These are Primary Agents that have selectively drawn upon cultural ideas to articulate their goals and develop an organization, making them more formal and institutionalized. For example, individuals of common concerns who have a vague and undefined concept that the human-land interrelationship is in a bad way are in no position to perform goal-directed action until they have developed or assimilated an ideology such as environmentalism that allows them to organize and articulate their concerns, as well as make knowledge claims (see also Fligstein 2001 for very similar proposals). Instrumentalist social power is thus largely to be found in social groups. Archer's specific formulation of Critical Realism, therefore, leaves individuals in a very weak position to effect purposeful structural change, unless they add their voice to an ideologically directed and defined collective of some kind. This is a distinct position from the commonly cited notion that ideology obscures true interests rather than providing direction.

Archer supplements agent and structure with culture, another field composed more exclusively of concepts and ideas that has a recursive relationship with structure, which is for her defined more as social organization.[11] Morphostasis (or the reproduction of structure) is most prominent if there is a single Corporate Agent that dominates both the cultural and social fields, as in the case of the church in the medieval education system (Archer 1996[1988]:274-287). Not only

---

[10] For those who must know, the Self is the continuous sense of individuality formed through practice, and is theorized by Archer primarily to demonstrate that individual personalities are not formed entirely by socialization. Her concept of the Person is less clear to me, and appears to indicate a more mature sense of individuality that is formed through socialization, practice, and 'nature'. The approach is clearly more mechanical than other treatments of similar issues (such as Moore 1994).

[11] The first half of Giddens' definition of structure – as rules and resources – seems to correspond better to Archer's culture. What Archer calls structure or society includes social organization and is precisely the area that is vaguely treated by Giddens.

were the teachers members of the clergy (roles in the social field), but they also provided the cultural logic for education (teaching future clergy). Primary Agents existed who differed with this view, but without a cultural ideology to articulate their complaints, they could not become Corporate Agents. Morphogenesis (or the transformation of structure) predominates with the proliferation of Corporate Agents providing avenues for some members of society to present and articulate their grievances and to strategically effect changes in structure.

I am hardly arguing that Critical Realism should be adopted in all its particulars. Both Structuration and Morphogenesis carry a heavy load of terminological baggage and analytical peculiarities, few of which are helpful for archaeology. Another bothersome element in Archer's formulation is the degree to which she relies upon old classic sociological studies such as those of Weber to conclude that pre-modern societies are excessively homogeneous (a single Corporate Agent) and more morphostatic, while modernity is characterized as inherently morphogenetic with numerous Corporate Agents (Archer 2000:270-282). Other sociologists, including Giddens and Bourdieu (see Beekman n.d.a.; A. Smith 2001), tend to make the same mistake of setting pre-modern societies off as tradition-bound and static in their analyses, and we should not feel obligated to retain all elements of these approaches. The approach as it exists in sociology retains a narrow view of social change, and the full infusion of these models with anthropology's broader experience has only begun.

Finally, Archer continues a tendency (followed by Giddens to a lesser extent) to reduce both individual or group motivation to competition and the pursuit of boundedly rational self-interest. Cooperation is an integral part of society as much as competition, but it has lost theoretical ground in recent years through its association with functionalist approaches. To be sure, human behaviour cannot be divided so simply into two opposing categories and competition can be the source from which cooperation springs (Conte and Gilbert 1995:6-8). But my point is that even boundedly rational self-interest is not the universal determinant for social behaviour. Groups can arise through reciprocity, stress-induced bonding (Freeman 1995), or individual psychological needs (e.g., trust – see Sztompka 2000). Vested interests are commonly set aside as an individual adopts another perspective to be liked or accepted or to avoid guilt, defers to others to avoid conflict, or even shamefully accepts that their own prior position was incorrect. Elites surely tempered their self-glorification with the emic sense, defined by their ideology, that they had responsibilities towards their subjects. These examples might be questioned, or even reinterpreted to make them again reducible to self-interest. But I do not think we want to do that. Our over-simplification of the sources of human agency is doing precisely what Clark (2000) saw as unhelpful in Darwinian approaches – the reduction of social explanation to innate human tendencies. Just as archaeologists in the 1960s and 1970s overextended their use of adaptation and function, the concept of individual self-interest was elevated to a causal force by archaeologists in the 1990s without consideration for how this characteristic differs between different cultures. We should consider making greater use of the literature on consciousness, the unconsciousness, and their roles in forming tendencies, motivation, and emotion (e.g., Ellis and Newton 2000).

Implications for Complex Systems Views on the Agent

Much of the previous discussion will surely resonate with students of complex systems, in particular the issues of emergence and the increased focus on relationships between agents. Some have even drawn a direct connection between complex systems and critical realism or related approaches (Baert and de Schampheleire 1988; Harvey and Reed 1994, 1996; Reed and Harvey 1992). Similarly, it should interest agency theorists to know that complex systems approaches have, in the various disciplines they have impacted, promoted a shift from laws to historical contingency, from unilinear or multilinear to nonlinear change, from closed to open systems, from homogeneous to heterogeneous agents, from top-down models to those based on self-organization among a system's parts, and from prediction to retrodictive explanation. While some look upon complexity as a 'Post-Modern Science', most practitioners see it as steering a course different from either subjectivism or mechanical objectivism.

Many complex systems researchers have made extensive use of computer modelling due to the amount of data and variables being analyzed, and the variant with which I am primarily occupied here is agent based computer simulation. Agent based models have typically oriented their studies towards demonstrating that a complex pattern, long thought to be the result of conscious design, instead emerged out of the uncoordinated actions of a myriad of agents 'from the bottom up', using a few simple rules (e.g., Castelfranchi and Werner 1994; Epstein and Axtell 1996; Gilbert and Conte 1995; Gilbert and Doran 1994; Gilbert and Troitzsch 1999; Kohler and Gumerman 2000; Langton 1988, 1994; Langton, et al. 1992; Nickles, et al. 2004; Suleiman, et al. 2000). The reliance upon a methodologically individualist logic is evident, but as the discussion up to this point has emphasized, this is only half the story when it comes to human societies (see also Hommon this volume). There appears to be some acceptance of the recursivity of structure and agency present among some of those who study chaos and complexity in social systems (Marion 1999:215-234), but this does not appear to be well developed in simulation work (see McGlade 2003 for related concerns about scale). If any constraining or enabling structural elements are present, they are typically programmed and fixed into the simulation architecture, and they are not allowed to emerge or change through agents' interaction. Several researchers (Gilbert and Troitzsch 1999:92-120; Lake 2005) have attempted to draw simulators' attention to this modelling gap, and Gilbert has specifically made reference to Structuration theory (Gilbert 1995). As he notes further, real humans are capable of examining emergent patterns in society and modifying their actions in various ways to strengthen or counter the pattern or offer alternatives, a uniquely human characteristic he (Gilbert 2000:365-366) calls Second Order Emergence.

If agent based computer simulation is to go forward in archaeology, it must overcome two growing perceptions – 1) that it is the instrument of strictly selectionist or evolutionary ecology-type approaches, in which agents are atomistic individuals with their own goals who do not communicate or otherwise interact with other agents; and 2) that it fails to deal with institutionalization and the top-down effects of political power. The first perception exists because of the frequent expedient recourse to modelling the transmission of social traits through genetic algorithms (Holland 1998) and the widespread use of rational actor models. In the same vein, there has been substantial interest in modelling fitness landscapes (e.g., Kauffman 1999), with less regard for the possible variety of measures of social success (Marion 1999:269). Theoretical work that shows how to model belief systems, symbolic thought, etc. seems very promising, but is infrequently applied to real case studies (exceptions include Doran, et al. 1994). There has also been an extraordinary amount of thought, ingenuity, and labour invested by some simulators (Dean, et al. 2000; Kohler, et al. 2000; Kohler and Van West 1996) into reconstructing environmental parameters without spending comparable effort on modelling the mechanisms of social and especially political interaction (see Lake 2000 for an exception). In fact, interaction *per se* has been sparse in the recent generation of social system simulations (see Johnson, et al. 2005 for the most cutting edge improvements here), which leads me to the second point.

Although several agent based computer simulations have been developed that model real modern or archaeological societies, simply using agents does not turn a simulation model into a complex system. Any agent based studies of which I am aware that are able to demonstrate the presence of dynamical chaos or even the less well-defined 'complex' behaviour have invariably incorporated the effects of one agent's behaviour upon another (Byrne 1998; various in Kiel and Elliott 1996; Lansing 2000). Lansing (2000:218) has shown how increasing the number of interacting individuals can effect a shift from paralyzing order to complexity to frantic chaos (see also Nowak, et al. 1995). It is, after all, the push and pull of different factors and the iterative feedback between variables or agents that creates nonlinear patterning. Agent based simulation simply cannot be equated with nonlinear systems.

It is important to point out that the debate over the kind of agent and agency to be modelled is to be found among Artificial Life simulators as well (e.g., Castelfranchi 1998; Macy 1998). Some analysts have a particularly impoverished concept of agency and treat it as purely stochastic (e.g., Saperstein 1997:112-113); it is unclear what such an approach is supposed to accomplish, as even ants act in a more patterned manner than this. More productively, Doran (2000:101) has noted that simulators have largely modelled societies as *egocentric*, in which individuals' goals, feelings, experiences, etc. are the focus of the model (although the various potential definitions of rationality possible within this category have been nicely discussed by Lake [2004]). This is clearly the dominant approach used to date, and there is some similarity to Giddens' and perhaps Archer's concepts of agency. Doran contrasts this with a *sociocentric* model, in which social roles are paramount. Individual agents would be programmed to maintain a memory of social situations, and how s/he is to behave in such contexts. Bourdieu's concept of practice seems more closely implicated here. Something akin to this is also implied in Lansing's Balinese model (Lansing and Kremer 1993), in which agents base their decisions upon the perceived successes of their neighbours.

This kind of approach requires a much more explicit recognition of the vast body of existing social theory, rather than falling back upon in-house methods of modelling agents derived from existing computational mechanisms. Schillo and colleagues (Schillo, et al. 2001) present a detailed discussion of how to model social systems as conceptualized by specific social theorists, and their suggestions bear little resemblance to the rule-following agents of better-known archaeological simulations. Some explicit attempts to model living societies using Giddens' Structuration theory (Sean Downey, personal communication 2002) and Bourdieu's habitus-field (Alam 2004; Schillo, et al. 2001) do exist, though they are primarily to be found among European researchers.

Clearly the autonomy of individual agents within a multi-agent system[12] is partly what is at issue (e.g., Nickles, et al. 2004), whether it is autonomy from other agents (e.g., Schillo and Fisher 2004) or in some cases the autonomy to ignore the social rules programmed within the simulation (Carabelea, et al. 2004). But rules should also be capable of varying between agents; this is what is implied by Bourdieu's concept of *habitus*, after all. To impose the same rules of a simulation onto all agents is a return to the old culture-historical monolithic view of culture that was rejected by archaeologists in the 1960s and 1970s. Perhaps what is required is an approach whereby the situational rules themselves 'emerge' through individuals' actions and 'stabilize' for particular groups of agents, instead of being programmed by the analyst. Sociocentric or egocentric behaviour could emerge along thresholds that differ among heterogeneous agents (Glance and Huberman 1993, 1994) or within certain social contexts, rather than by designing agents to deterministically interact with x number of neighbours or to pursue only one or two specific programmed goals. The only rules that should be fixed within a simulation should be the most basic biological imperatives common to all humans, while any Giddensian structural rules and resources must emerge through agents' actions. A still more radical solution would be to abandon the idea of modelling individual agents at all, allowing agents themselves to emerge from below as bundles of fluctuating practices formed through repeated interactions with others (Freeman 1995 provides a stimulating model for this kind of approach).

Solving the problem of what kinds of agents to use does not get us past the problem of collective entities in agent based simulations. Even should a large number of interacting agents be placed upon a simulated landscape, they will predictably never form a society because while emergent properties or patterns may occur, they are not being given sufficient cohesiveness to either reflect back upon and influence agents' behaviour or to serve as a new staging point for further

---

[12] I recognize that a multi-agent system is not the same thing as an agent based simulation. The former is more oriented towards the design of cognitive models, but some of the problems they have been addressing are of parallel interest.

activity. Unless collective agents are capable of coalescing into forms partly disarticulated from agents' actions, it is hard to conceive how new rules or institutions or truly *political* activity could impact upon agents. The problem is parallel to that which hangs over Structuration theory – how can we acknowledge the important effects of structure and collective agency while still retaining an emphasis upon the theoretically important element of individual action?

Giddens warns us against the reification of structure, and is so concerned about this as a tendency among analysts that he avoids the issue of emergence and collective agency. But it seems as if something akin to this is the process we are observing as collectivities or organizations *institutionalize* (well argued in Sawyer 2004). Consider for a moment the example of a political faction or party, which has many constituent members. That party is a collective agent, typically with an associated ideology (*a la* Archer 1996[1988]) that is *always* an imperfect and hence decentred rendition of the interests of the faction's membership. That ideology is a sufficiently simplified model of reality to attract and incorporate agents and allow them to articulate their concerns about a certain range of issues. This simplified view of reality both constrains individuals by providing them with a limited number of choices as to what views to hold and what parties to support, but it enables in that it provides opportunities and paths to express their views. It tremendously simplifies the decision-making process for individuals by allowing them to articulate their position with a small number of decisions (e.g., what faction to join), instead of personally expressing themselves on each issue in all its nuances. It brings 'order' to chaos by imposing a kind of tunnel vision, winnowing away a nearly infinite range of responses into a more limited range of resulting patterns, which incidentally stabilizes that collective agent and actually promotes its continued existence. This certainly does not preclude new ideologies or changes in the old ones, but does provide the context within which such innovation can occur. It also does not preclude resistance by others within the group; it may in fact explain it. Resistance to authority may well occur precisely because of that process of trimming down the variation in positions and goals.

Collective agents, most likely stabilized through a simplifying ideology or worldview, are what Harvey and Reed (1996) call symmetry-breaking. They dampen the chaotic behaviour that grows with the size of the interacting group (Lansing 2000:217-219) and create a shifting platform for further hierarchical or heterarchical (Crumley this volume) developments. A true populist democracy would likely exhibit chaotic behaviour, but a representative republic utilizing a multi-party system introduces a separate layer of agency and goal-directed action partly removed from individual psychological dispositions (Richards 1997:112-113; Schofield 1995). The numerous different positions of individuals are instead aggregated, redirected, simplified, or ignored, and the actions of individuals are to a great extent contained by the collectivities they access to effect change.[13] These

[13] Network analysis (e.g., Emirbayer and Goodwin 1994; Schweizer 1997) seems to be on a parallel track here, and promises useful approaches to analysis and to map the form and extent of some kinds of collectives (see Bentley 2003a in archaeology).

examples are each associated with ideologies that both imperfectly render and justify the structure while also guiding and limiting individual action.

These collective agents make further differentiation possible. The formation of higher order political units does not appear to occur until there has already been some lower order stabilization. The European Union could only form as a very high order collective agent after its member polities (old style nation states and themselves collective agents) had long since stabilized. This is certainly not to say that they were in some kind of homeostatic equilibrium, but the component political institutions were perceived as likely to be there for some time to come. These states had, in fact, experienced new relational (competitive global economics) and ideological (anti-nationalistic) shifts that preceded the move towards a new attractor. On the other hand, the consolidation of the United Nations as a world government with significant power continues to be hampered not only due to resistant nation states (such as the United States) but also because other potential components (such as the Near East, the Balkans, etc.) are so fragmentary as to prevent incorporation.

Of course, there is volatility present in any corporate agent partly because its individual members do not have a single identity but embody potentially numerous cross-cutting identities (gender, age, class, ethnicity, religion, etc.). Ethnic groups, tightly knit Midwestern religious communities, and some other groups tend to combine multiple isomorphic identities such as class, religion, and ethnicity, making them hard to leave but comparatively effective representations of their members. When these groups are forced to incorporate an increased diversity of social actors, the potential increases that member agents will choose to associate themselves with some other group because they feel the group no longer represents their interests. But those collectives that can successfully manage and represent the variation within them may be more likely to hold together.

Clearly, none of these collectivities exists in anything close to an equilibrium or homeostatic state. Rather, each is a moving target that continues to shift in its organization, ideology, and membership. For this reason I suggest that we instead speak of 'stability' as Prigogine does (Nicolis and Prigogine 1977; Prigogine 1980; Prigogine and Stengers 1984). This is essentially an attractor (Marion 1999:235-240), but complex attractors are said to be different from chaotic attractors in that they retain sufficient stability to allow 'memory' (Marion 1999:73-74). The institutional memory stored in writing, other symbols, material culture, or spatial arrangements are what prevent complete decomposability of complex systems far from equilibrium, just as individual memory makes it impossible for humans to rewind the tape and act as if certain events or periods in their life had not occurred. Even when a political or economic collapse occurs, that institutional memory prevents society from reverting to a previous state and instead defines a 'new' state. The slate is never wiped clean and the arrow of time always points forwards. These observations allow us to speak of collective agents as real social entities that are observable, capable of being analyzed, and possessed of emergent properties that impact individual agents. The shifting actions of individuals will continue to alter these institutions to varying degrees, although increased dissociation from individuals may well explain complete institutional collapse.

Once formed, institutions transcend individual relations and negotiation to allow top-down and truly political action. Hommon (this volume) argues that stratified control systems are a feature that distinguishes some social systems from other complex systems, and that this feature has not been adequately approached by many complex systems theorists (see also Khalil 1995). Some attempts to create this type of modelling environment exist among European scholars. Carley (1997) and Egidi and Marengo (1995) have developed stratified simulations in which agents at the higher level are endowed with strategic options as managers that are not available to those agents engaged at the operational level. The holonic multi-agent system (Fischer, et al. 2003; Knabe, et al. 2003; Schillo and Fischer 2004, n.d.; Schillo, et al. 2002) is one in which there are agents that are simultaneously composed of other agents. Collective agency can thus form through the actions of individuals, but this does not mean that these higher order units are immutable. Some simulations explicitly characterize organizations as composed of individual agents, and the focus is upon how the organization (implicitly characterized as an agent with its own goals) tries to convince members to act in a certain manner without the ability to force compliance; one author does this through defining expected behaviour and imposing sanctions on deviations from that behaviour (Pacheco 2004). All this is a far cry from adaptive, optimizing, or even satisficing models of human behaviour – the approaches described here are explicit attempts to deal with power as it is exerted over other individuals, which constitutes a completely different process from that used for strictly bottom-up simulations.

What is ironic here about these creative attempts is that it should be the social scientists properly cautioning the computer modellers how to correctly consider social theory when doing computer simulation. Instead, in the absence of participation by social theorists, the simulators have taken it upon themselves to incorporate theoretical models. They appear, in fact, to be more sophisticated than the simulations designed by many anthropologists and archaeologists. This needs to change.

To conclude this argument, agency approaches are highly important and necessary for shifting the level of explanatory narrative to those who truly make decisions in a society. This means that we need to conceptualize agents so as to fully appreciate their folly, rationality, short-sightedness, inaction, perspicacity, will to power, and uncontrolled desires. It also means that institutions hold important positions as filters for social rules, unparalleled mechanisms for instrumental action, repositories of memory, tools for protecting one's social position, and as sources of status and identity for members. Institutions possess a degree of independence from social action that demands that they receive due attention as more than just the aggregated actions of members. Simulation research is making inroads towards social modelling, but many models of past societies have aligned themselves with a narrow theoretical orientation that models the human animal, not the social human. Future simulators need to be more aggressive and self-aware in their use of different theoretical approaches, and recognize that modelling 'from the bottom up' is not neutral empiricism, but the unconscious adoption of a dated paradigm inappropriate for human society.

Pre-Columbian Western Mexico

The preceding discussion has highlighted the emergence of collective agency and the need for different analytical rules at different scales. No one example is likely to effectively illustrate all the points I have discussed, but I would like to selectively examine some of these issues in Pre-Columbian Mesoamerica. I will discuss examples using my work in central Jalisco, Mexico, namely the appearance of, and the relationships between corporate descent groups. There are some similarities to Mark Lehner's discussion of scale in Pharaonic Egypt (Lehner 2000), but with more emphasis upon phase transitions, or differences across scales (cf. McGlade 2003), as we might expect if we espouse the idea of emergence.

The region in question is comprised of a series of linked highland valleys and lake basins surrounding the Tequila volcano in Jalisco, western Mexico (Figure 4.1). The society located here during the Late Formative and Classic periods (ca. 200 B.C.-A.D. 500/600) presents evidence for the expression of social inequalities through variation in mortuary ritual, quantities and qualities of grave goods, a complex settlement hierarchy, monumental public architecture, and irrigated raised fields for intensive maize agriculture (e.g., Beekman 2000; Stuart 2003; Weigand 1985). While clearly a society with a significant degree of hierarchy, many questions remain unanswered due largely to the dearth of excavation at the local level. The current discussion draws partly upon my recent excavations at the sites of Llano Grande and Navajas, from opposing ends of the Tequila valleys.

Recent field research has documented the presence of corporate descent groups as emically significant social units within this society. Deep shaft and chamber tombs appear to be used for successive family interments (Figure 4.2; Beekman 2000; Pickering and Cabrero 1998; Ramos and López M. 1996), and these same family groups are considered the likely builders and custodians of platform temples sometimes built directly atop these tombs (Beekman n.d.b.; Butterwick 2004). Individuals clearly exist, of course, and the methods by which they affiliate with these corporate groups may be various. There is biological evidence for kinship affiliation among the members of these groups (Pickering and Cabrero 1998), but of course there is nothing automatic about biological ties for defining social groups. Kinship may be variously defined, and I consider ties of affiliation and non-biological kinship potentially relevant for defining group identity as well (Beekman n.d.b.). The question is, why did these social groups exist at all?

When I first wrote on the topic (Beekman 2000), I attempted to reconstruct a plausible rationale for why unaligned social actors other than those linked by descent might desire membership in such a social group. Following then current approaches to agency, I focused on competition between elites for power and

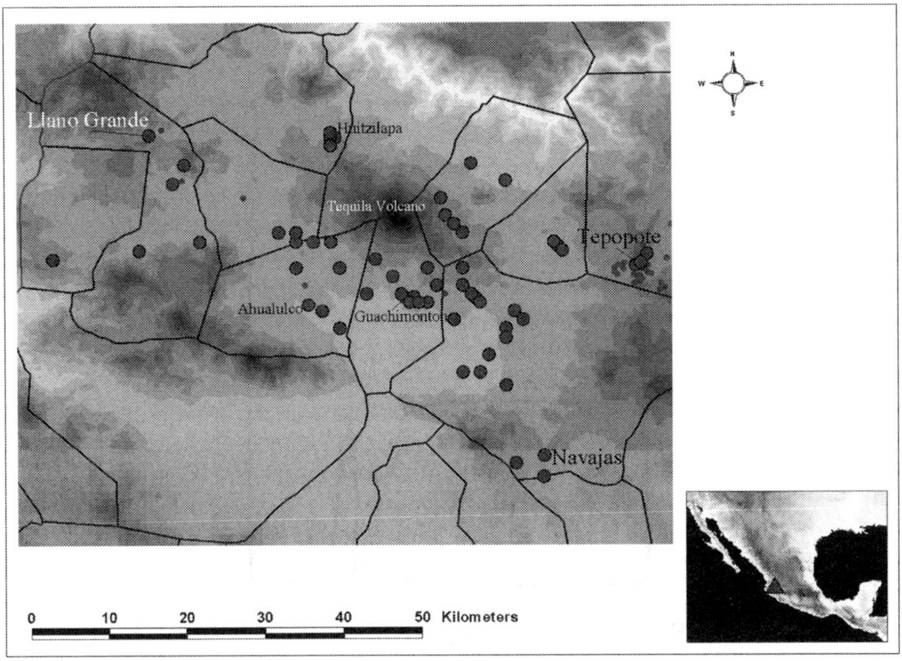

Figure 4.1 Map of the Tequila valleys of central Jalisco, Mexico

labour, and on those non-elites who might seek the security of affiliation with the more successfully competing elites. As competition continued, however, the repercussions of *not* supporting one contender or another might become too severe to avoid throwing one's hand in, regardless of whether one agreed with the broader social rules that allowed or encouraged that competition. The logic is of course entirely circular. Elites compete for non-elite labour, while non-elites are looking for elites to affiliate with. It is not clear why any of it is happening, unless we posit the presence of inherently competitive individuals within any given population (*a la* Clark and Blake 1994) or, following common chaos approaches, we might emphasize how small (and perhaps unpredictable) variations in initial conditions gave competitive advantages to some individuals over others, that were eventually magnified into more significant social variation.

More recently there has developed another way out of this impasse, inspired partly by Archer's argument that corporate agents form around an ideology or 'mission statement' that define them. Corporate groups usually define access to some kind of capital resource. I have argued that our social groups were the vehicles for access to different kinds of corporate property, most likely positions in a political/religious hierarchy (for elites) and land (for non-elites) to judge from the locations where the ancestral tombs of each are found (Beekman n.d.b.). Non-elite

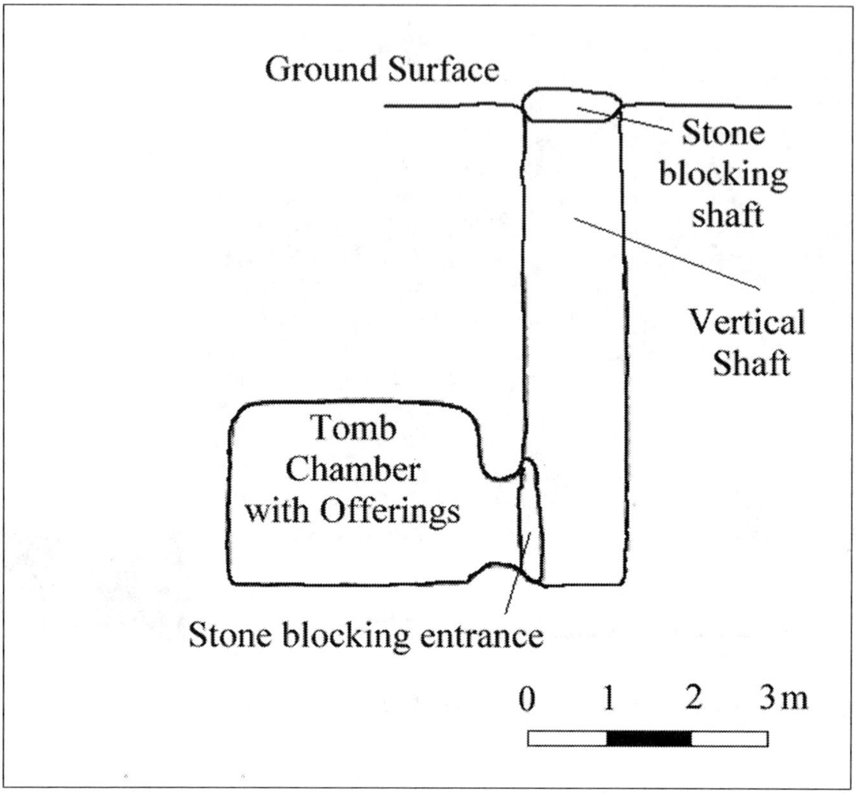

Figure 4.2 Example shaft and chamber tomb used for elite interments throughout western Mexico

tombs cluster together into cemeteries in the rural areas. Elite tombs are found beneath architecture, especially the platform temples mentioned earlier.

Presumably there is no need to specify access to property unless there is a limited quantity of the resource available, or unless the quality of the resource is highly variable. The former likely describes the case with political and religious positions, while the latter probably describes the variable quality of land in the region. Explaining the very existence of our social groups by reference to capital of one sort or another more successfully welds together the disparate arguments about why anyone would want to join them. It also clarifies that the competition among elites to obtain followers may not have been about linking unaligned individuals to a group, but rather it may have been a mechanism whereby non-elite lineages became attached to elite lineages, or food-producers to non-food-producers.

Of course, all these proposed motivations are perhaps more apparent to me than they might have been to the actual members of this society. We should not discount more visceral and less rational reasons for desiring group membership, whether we are talking about the very personal and psychological gains to be had from joining friends in their decisions to join group A or B, fears of being left behind, feelings of duty or loyalty for past deeds, hostility towards particular groups, etc. Many of these non-rational or irrational reasons for participation seem more closely derived from interaction with other members of society, than upon acting to advance one's personal situation. Indeed, group formation may even require suppression of individual, Darwinian, motives.

Let us move on to how these different social groups interacted with others. As I have already implied, there is reasonable evidence that they were in social competition over access to followers. Significant quantities of exotic and imported goods appear within the same ancestral shaft and chamber tombs mentioned above (e.g., Cabrero G. and López C. 1997; Ramos and López M. 1996), and contemporary ceramic dioramas depict what appear to be burial processions, suggesting public performance in which one's wealth and connections are openly displayed as a tool by which elites competed for followers' labour (Beekman 2000). Competition may have become violent to judge from the presence of ceramic representations of prisoners (Townsend 1998: Figures 5, 6, 8-11), warriors (Kan, et al. 1970: Frontispiece, Figures 12, 64, 127, 128; Von Winning 1969: Figures 74, 140-141), and battle scene dioramas (Von Winning 1969: Figure 154), but this could easily represent warfare between polities and not our small social groups. If competition between groups was over followers and labour as proposed here, this might be seen as a comparatively pragmatic goal compared to the psychological factors advanced earlier as individual motivations, and exemplifies what I mean by institutions becoming distanced from individual dispositions.

Yet despite this evidence, these corporate groups are brought together in highly choreographed arrangements of public architecture that imply a very different sort of relationship and yet another level of identity. The temple platforms introduced above were built exclusively in symmetrical circular arrangements, composed of eight (or occasionally ten or twelve) platforms, that all faced in onto a common circular patio (Figure 4.3; Weigand 1985). With the addition of a circular altar or pyramid occupying the centre position in the patio, these complexes were overwhelmingly the primary form of public architecture in the Tequila valleys of central Jalisco from at least 200 B.C. to A.D. 500/600. While the material evidence for competition between the groups lies in the tombs beneath the platforms, within the public space of the surface architecture there are few elaborate artefacts and the evidence for competition is much more muted (Beekman, et al. in prep.; Tyndall n.d.). Archaeological evidence and ethnographic parallels with the Huichol and other indigenous groups of western Mexico too detailed to go into here (see Beekman 2003a, 2003b, n.d.b.) suggest that the different corporate groups each held complementary roles within the larger schema of annual agricultural ritual, suggesting significant responsibility for community well-being. No one group was responsible for ensuring success in subsistence endeavours and no clear 'leaders'

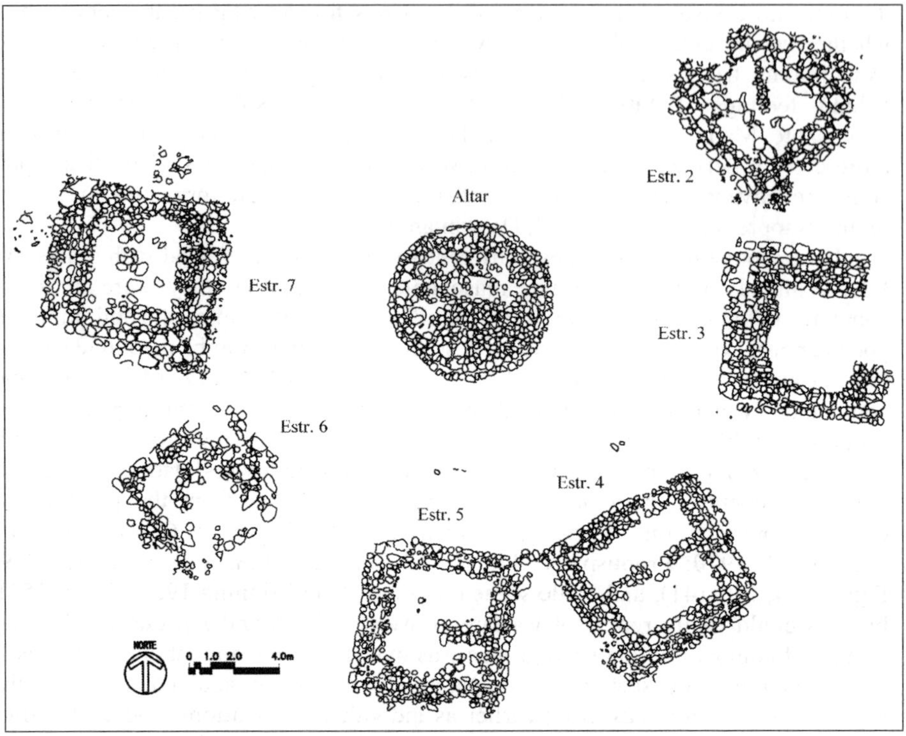

Figure 4.3 Circle 5, Navajas, Jalisco, Mexico

except possibly heads of the different lineages are portrayed in imagery (e.g., Beekman 2003a: Figure 13; cf. Butterwick 2004). Participation in group ritual in the formation of this higher order social unit is a good example of what Freeman (1995) describes as the basis for group formation – the bonding that takes place in socially stressful activities such as ritual initiation (see also Stone, this volume).

The ideology associated with the corporate agent (in Archer's sense) represented by the eight lineage groups interacting within the public architecture both define these groups as elite, and unite them with commonly held interests in opposition to non-elites. Elites may have held a communal orientation, such that they held a shared ethos as the individuals responsible for maintaining community well-being – this is actually implied by their role in agricultural ritual. Of course these groups may also have acted in common self-interest to maintain their privileges, but it seems clear that social rules assigned those privileges not to individuals, but to the corporate body. Maintaining access to elite corporate property might involve both supporting social rules that define elite rights and maintaining membership within the elite category. Non-elite individuals or lineages seeking to gain access to corporate property can pursue two strategies –

joining with an elite group to obtain rights (supporting structural rules defining elites) or attempting to change the rules so that access was not limited to these elite groups.

All this may seem excessively speculative for an archaeological case study. But this exploration of possibilities has served to help clarify that social actors are rarely interacting directly with the rules and resources of structure, but with the way in which structure has been instantiated through collectives. Both the individual corporate groups and the larger collective of eight interacting corporate groups are each defined by an associated ideology. The former defines access to capital while the latter is defined by the groups' complementary and combined roles in religious ritual. With or without claiming that the elite corporate groups have become a single social actor, they are nonetheless the face which structure presents to individual agents.

Assuming that this is a legitimate reconstruction, why and how would one go about simulating it? Individual action and motivation are deduced from aggregated archaeological data, but rarely do archaeologists reverse that process and attempt to test whether the proposed individual actions would truly result in the larger pattern originally observed in the material record. In order to be fair, I shall take an example from my own research. I had proposed (Beekman 2000) that the competition between elites for followers through public performance and display of exotic imports may have led to the emergence of a core-periphery pattern. The distinctive circular temple arrangements in central Jalisco reach their greatest expression in that region, but smaller derivative examples are found across the adjoining states along paths that suggest trade routes towards sources of exotic rare resources. If social competition in central Jalisco relied upon the use and interment of such exotic goods, then local strategies between our corporate groups might well have led to the emergence of this regional economic network.

Testing this proposition could involve modelling competitive aggrandizers at the local level and the quantities and rates of exotic goods that might be used. The results could help determine whether the agent based narrative grossly over-predicted or under-predicted the volume of material exchanged, as measured against materials excavated from final contexts, materials quarried from source areas, etc. Further, in keeping with the archaeological evidence, we would have to model competitive mortuary display and cooperative temple ritual using more of what Doran called a sociocentric approach. That is, the behaviours are not inherent in the individual, but are cued *to* individuals depending on the social situation, so that they learn with time that certain locations or contexts are appropriate for certain behaviours and others are not.

Simulation does not replace other traditional means of data collection or analysis, certainly not the reconstruction of ideological content as in this example, but it does help to introduce dynamism to a field reliant on a static archaeological record. Field, laboratory, or other background research still provide the data that set the parameters for any simulation. Just as the MASS project (Wilkinson, this volume) has made use of cuneiform texts to construct the raw structure of their simulation, field and iconographic studies will be indispensable to constructing any viable simulation. Interpretations such as those discussed above (the opposition of elite vs. commoner lineages, etc.) still need to be made, but can be made more effectively when there are additional data to clarify the picture.

What simulation is especially likely to do for agent based studies in archaeology is require theorists to specify relationships between rules and action, action and reproduction/transformation of society, etc. to a far greater degree than is currently accepted. To judge from the literature, this is a problem that particularly plagues studies of agency. Verbal modelling is highly forgiving, and allows us to gloss over missing information or dubious assumptions in our narratives. Simulation mercilessly requires specification. If one can specify relationships (e.g., 'Elite tombs are found beneath public architecture') or assumptions (e.g., 'elites have more information on which to base decisions than commoners'), they can probably be simulated – if one cannot, then neither verbal nor computer modelling will fill the gap. A deft writer can obscure the weaknesses of their argument – a deft modeller cannot. A verbal model requires the writer to concentrate on narrative and the most effective way to present their position, and one can easily miss an important weakness in one's argument as a result. A computer simulation forces the designer to confront issues and assumptions to an alarming degree. Once created, a simulation is also a framework that allows other analysts to tinker and alter those assumptions in line with their own proposals, and evaluate other potential explanations.

A final example will suffice before I lay this Procrustean argument to rest. We have discussed very detailed examples of agency and group formation up to the scale of the integrated multi-component architectural circles. The site of Navajas is an example of a medium sized settlement within the Tequila valleys and is currently dated to 50 B.C. to A.D. 200 (Figure 4.4). As the map of its central collection of architecture shows, there are four circles within a stone's throw of one another, and another five or more just beyond the bounds of this map. They vary significantly in volume and construction effort, and their relationship remains unclear. Are they contemporary? Are they hierarchically nested? Do they share power heterarchically (Crumley, this volume), perhaps sequentially? Questions such as these come much more strongly to the fore with the perspective outlined here. Much remains to be explained, but the group relations that we have discussed up to this point represent only the very smallest scales of group organization within this society. Emic definitions of community await clarification, not to mention the relationship between ceremonial centres across the Pre-Columbian landscape. The Tequila valleys as a whole may even have been temporarily unified under a single political administration that directed the establishment of a centralized boundary monitoring strategy (Beekman n.d.a.) as an emergent property unique to that scale. What role individuals played in cross-polity political relations remains to be seen, but models that ignore the potential agency of collective groups in favour of individuals risk the label of reductionism.

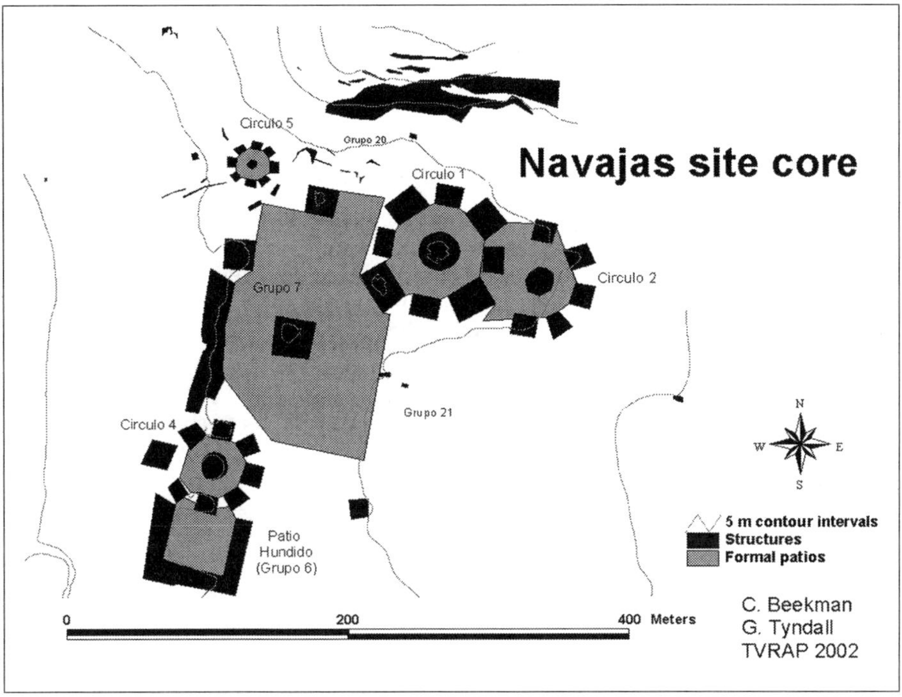

Figure 4.4 Mapped portion of the ceremonial centre of Navajas

Conclusions

The application of the principles described in the first part to the study of the archaeology of Jalisco is rough and incomplete. But I think that it helps to bring local research into focus. There is a need for a stratified approach that does not reduce activity at all scales to the same common denominator of individual agency. Collectivities are important units of analysis and corporate descent groups form significant and stable corporate agents representing their members in the pursuit of certain goals over a very long period of time. The ideologies that accompanied the formation of these social groups guided them through idealized definitions of the relationship between one group and another. Further metaphors probably defined the wider community, and specified the proper roles between outlying centres and primary centres. The possibility of a single valley-wide political system is still under examination. The point here is that at each of these scales new rules emerge that define the relationships between family members, factions, communities, or 'polities' (cf. Lehner 2000) and, indeed, between the components across different scales (T. Smith 1997:59). These rules will be relevant within a limited range of contexts that may or may not include the scale of individuals. But the relevant structural factors that constrained, enabled, or otherwise defined the rules of interaction between those individuals emerged along with collectivities intermediate to structure and agency.

Acknowledgements – A shorter version of this chapter was presented as *Critiques of Giddens' Structuration theory and the implications for complex systems views of agency* at the 101st Annual Meeting of the American Anthropological Association in 2002. I would like to acknowledge the continuing support of the National Science Foundation for past and ongoing work at the sites of Llano Grande and Navajas. The Foundation for Ancient Mesoamerican Studies, Inc. and the University of Colorado have also supported this research. The Instituto Nacional de Antropología e Historia has graciously provided field permits. This work 'emerges' from continuing collaborative research with William Baden.

Chapter 5

Factional Formation and Community Dynamics in Middle-Range Societies

Tammy Stone
University of Colorado at Denver and Health Sciences Center

The archaeological landscape is populated with numerous sites associated with pre-state agricultural groups. When discussing the formation of these communities, archaeologists frequently talk about daughter villages 'budding off', or parent villages fissioning, without detailed attention to the internal dynamics involved. This chapter examines these internal processes and offers a model with which to understand village fissioning. It then examines the historically recorded split of the Hopi Village of Orayvi[1] in 1906 to illustrate this model.

Communities in middle range societies, even those that appear relatively stable for long periods of time, are not internally homogeneous (see Brumfiel 1992 and Brumfiel and Fox 1994 for a similar argument for state level societies). Rather, quasi-group formation and factionalism are constant as individuals and groups experiment with organizations in their attempts to achieve their own personal goals. These groups can be based on any number of shared relationships, including kin groups which become increasingly important with the formalization of sequential hierarchies associated with larger communities (Hays 1993; Johnson 1982); ritual affiliation (Brandt 1994; Saitta 1991; Stone 1994, 1999); and competing interest groups in the political arena (Stone and Howell 1994).

An enduring question in anthropology regarding middle range societies, especially middle range groups living in large communities year round, centres on issues of the tensions between subgroups in the community, regardless of the basis of their formation, and the balancing of competition and cooperation between subgroups at various scales (individuals, households, corporate groups, ethnic groups, competing interest groups, etc.). Nonlinear modelling is a particularly useful approach in studying this issue because of its attention to the interplay of multiple scales, its sensitivity to event-specific (i.e., initial) conditions, and its concentration on variability and change within the system, as well as perturbations from outside.

---

[1] Spelling of Hopi personal names and place names follows those set out in the Hopi Dictionary Project (1997).

Factionalism and Quasi-Group Formation

In middle range societies – where there is still some fluidity in power structures in at least some portions of the society – individuals and aggregates of individuals at a variety of scales frequently move back and forth between cooperative and competitive relationships with others for a variety of economic and social reasons (Brumfiel 1992; Kohler and Van West 1996; Stone 1994, 1999). These fluid alliances in political and economic action have been usefully characterized as quasi-group formation (Mayer 1977; Nicholas 1977; Salisbury and Silverman 1977; Stone and Howell 1994). Quasi-groups are characterized by a semi-fluid (or quasi) membership. That is, there is frequently a core group of individuals who share a position in the societal power structure, and, therefore, often share similar goals regarding political action, at least on some sets of issues. On any one issue, the total membership of the group can vary considerably as these core individuals attempt to attract additional followers to their cause. It is important to stress that this is not a typological approach, but rather a very dynamic one for at least three reasons. First, the core group may not maintain its cohesiveness on all issues. That is, cooperating core individuals on one set of issues may be in competition and may, in fact, be followers rather than core members of a faction on other issues. Second, it is dynamic because of the highly fluid nature of the additional followers that the core individuals can attract on any given issue. Finally, it is characterized by variable results of factional fights, reifying the existing socio-political structure sometimes and resulting in either a fissioning of the community and/or the formation of a new socio-political structure in others (Salisbury and Silverman 1977).

Whereas quasi-group formation as a constant process in pre-stratified societies has been accepted for almost thirty years (Schmidt, et al. 1977; Silverman and Salisbury 1977), the reasons why this process occurs are considerably more controversial. I argue here that the reasons for quasi-group formation, as well as the reasons the impact of this process varies so dramatically from situation to situation, can be most profitably understood when viewed through the lens of nonlinear modelling and the importance of initial conditions. Nonlinear modelling concentrates on multiple scales of analysis (Spencer-Wood 2000), variation in individual behaviour, and sub-system interaction (Allen 1982, 1989, 1994, 2001; Clark, et al. 1995) as crucial factors in system change. Of particular importance in understanding social relations in general and factionalism in particular is the distribution of information. As used here, information is defined in its largest sense and includes not only knowledge of resource distribution and ownership, but, more importantly

> knowledge of social structure including social relations with both kin and non-kin members and the rights and obligations that accompany them; group membership and identity; the political, social and economic organization and the integration of these subsystems within the larger social system; relationships and contexts for interaction

with outside groups; and mechanisms for the coordination of activity on a group [and sub-group] level. (Stone 1994:11)

Information Nodes and Social Organization

Van der Leeuw (1981b; Van der Leeuw and McGlade 1997b) and Stone (1994) have argued that aggregated communities/towns represent communication nodes on the landscape. Due to their concentrated nature, these nodes represent disruptions in the equal spread of information across the landscape. These nodes serve as attractors (or hills of high probability) on a probability landscape (Allen 1994). That is, in making decisions about participating in aggregated communities, individuals and groups of individuals are drawn, or attracted, to these concentrations of information at a greater rate than areas lacking concentrations of information. The formation of information nodes on the landscape has profound affects on the sociopolitical organization of both communities and regions which must accommodate and process the increasingly concentrated information.

Social institutions (be they formal or informal) are constructed by individuals and groups of individuals to aid the flow and processing of information (Van der Leeuw 1981b; Stone 1999). Changes in the amount or kind of information thus have major repercussions for the social institutions through which it flows and the manner in which individuals and groups of individuals interact. Further, the complexity of the social organizations that structure and process the information increases as more information regarding cooperation and competition on a regional level is concentrated in particular locations on the landscape. These changes, in turn, affect the distribution of information. Specifically, this increased intensity in information processing has a centripetal force (Root 1983; Stone 1994) which pulls more information at an ever-increasing rate into a set location on the landscape. This centripetal force leads to an information 'vortex' (Van der Leeuw 1981b) in which information flow increases as individuals concentrate at these locations to ensure access to the information. As information becomes increasingly concentrated at the centre of this 'vortex', it is pulled away from outlying areas. Failure to participate in the information processing social structures means being cut off from important sources of regional information. As a result, individuals, households, and other social groups of varying scales concentrate themselves spatially at these information nodes and aggregated communities/towns develop (Stone 1994; Van der Leeuw and McGlade 1997b).

Quasi-group formation within communities can be viewed in a similar manner on a smaller spatial and social scale (community rather than region). That is, the interaction of the core members of the quasi-group can be conceived of as a concentration of information within a community wide social landscape. Just as there are multiple nodes on the landscape when aggregated communities are viewed regionally, there are also multiple nodes within the community when quasi-groups are considered. That is, quasi-groups do not operate in isolation but in opposition to one or more other groups within the community – hence their tie to factionalism (Stone and Howell 1994) – and they can be examined in terms of information vortices. To understand this phenomenon, we switch our area of concentration from regional distributions of communities to social structures within communities.

Information processing occurs in every social interaction (Rothenbuhler 1998), though we can see it and its impact on cooperation and competition most clearly in the archaeological record of middle range societies in the ritual sphere (Stone 1999). This is true because ritual is, at its heart, a form of voluntary collective action and performance, as well as a way of thinking and interacting (Rothenbuhler 1998). Further, ritual is particularly helpful as a medium for understanding quasi-group formation and action because it is 'performed in social situations and structured by social phenomena, they [rituals] ... have social meanings. They refer to relationships and social position ... they are a way of indicating and embarking on socially oriented intentions' (Rothenbuhler 1998:13; see also Robbins 2001). Finally, ritual is referential in two directions: backward 'to the social order and the culture in which the ritual is embedded and forward to the order the participants envision' (Rothenbuhler 1998:63). It is this potential for projection forward within action tied to the community and past, as well as the social performance of ritual, that makes ritual so important for quasi-group action and recruitment of individuals to the cause of the quasi-group by the core members in a factional fight. Thus, examining shifts in architectural structures and paraphernalia associated with ritual in the archaeological record can help us to identify and understand quasi-group emergence/formation in factional disputes within communities (Saitta 1991, 1994; Stone 1999, 2002).

The question remains, however, why would an individual or group of individuals, choose to cooperate with one quasi-group rather than another (i.e., be attracted to one information node rather than another operating within the same community)? In other words, why are the core members of quasi-groups able to attract sufficient followers relative to other factions to either ensure stability within the community or to institute change? The tie between the uneven distribution of information and political action within communities has long been recognized (Crick 1982). In ethnographic settings as disparate as the American Southwest (Brandt 1994), West Africa (Murphy 1980), and Australia (Campbell 1978), it has been demonstrated repeatedly that the control of information, particularly information used and distributed in ritual settings, is an important source of political power in middle range societies. I argue the relationship between the accumulation of information and quasi-group formation exists because individuals need access to information to pursue their personal goals. Thus the abilities of core members of quasi-groups to attract and maintain followers is tied to their ability to manipulate and concentrate information through ritual, thus forming information vortices within a community. The ability to succeed in this action is tied directly to positive and negative feedbacks within the system, the initial conditions of the conflict, and the closeness of the system to criticality (Bentley and Maschner 2001; McGlade and Van der Leeuw 1997) – issues for which nonlinear modelling is particularly helpful.

That is, whether or not core individuals within quasi-groups can attract followers, and create a powerful faction, is dependent on the action and decision making of the followers and how the actions of the faction will affect their access to information needed for the attainment of their own personal goals (Brumfiel 1992; Clark and Blake 1994). However, this is not a one-dimensional process in which individuals weigh cost/benefit ratios of a particular action in the absence of consideration of the actions of others (Allen 2001). To begin with, we must remember that individuals do not live their everyday lives in isolation from others so that the only consideration is what is the most 'efficient' means of obtaining a set goal (Stone 2003). Second, we must remember that personal goals are not the same for everyone because they are conditioned in large part by the individual's position in the existing social and political structure relative to others (Archer 2000; Bogucki 2000; Stone 2003). Kin relations, gender, age, ethnicity, past political alliances (to name only a few) affect an individual's view of what goals are realistic and even desirable. Because no two individuals have exactly the same life history and position within the social structure, no two individuals' goals are exactly the same. However, goals may overlap on certain issues for individuals in similar positions within the society (Archer 2000). Thus, while an individual may support a core member of a quasi-group on one issue in a political dispute because it enhances the probability of achieving some goal they share, s/he probably will not on all. This is further complicated by the fact that whether an individual chooses to affiliate her/himself with a quasi-group or faction in any one situation is dependent not only on highly variable personal goals but on her/his perception of the stability of the system (defined as facilitating the access to information needed to achieve personal goals) and probability of success at any one time. In other words, individuals make decisions reflexively within the historically bound constraints in which they live their everyday lives (Giddens 1991). Combined, these factors add a degree of uncertainty to the recruitment activities of quasi-group core members. Therefore, the initial conditions of a factional fight and the criticality, or stability, of the system have great influence on the ultimate outcome (cf. Bak 1994; Spencer-Wood 2000).

I have argued elsewhere (Stone 1999) that perception is one of the most important factors affecting both the formation of a powerful political faction from a relatively fluid quasi-group and the success of a faction in challenging the status quo. If individuals perceive the system as stable (i.e., facilitating the access to information needed to achieve the varying personal goals of the participants), they are less likely to heed a call for change on the part of the core members of a quasi-group (Stone 1999). In this case, negative feedback mechanisms in the system will dampen change and the existing socio-political, and therefore, information processing structures, will be reified. Conversely, if individuals view the system as unstable (i.e., hindering their access to information needed to achieve their personal goals), they may be willing to participate in a quasi-group that advocates challenging the status quo and experimenting with the existing socio-political system (Gemmill and Smith 1985; Stone 1999; Van der Leeuw 1981b). If this occurs, a series of positive feedback mechanisms will increasingly concentrate information in the locus of the challenging faction at the expense of other factions within the system, resulting in an information vortex. This in turn will precipitate changes in the organizational structures which process that information, eventually resulting in the arrival at a bifurcation point (Bintliff 1997; Van der Leeuw 1981b) where either a series of new organizational structures arise within the village or the village fissions.

The process by which perceptions of the situation shift is iterative and directly tied to the closeness of the system to criticality. That is, there are always individuals in any society that experiment with social relations and institutions (i.e., they participate in non-average behaviour). These are the core members of quasi-groups seeking change. In terms of nonlinear modelling, there is a constant and 'on-going dialogue between average and non-average behaviour within the system' (Allen 2001:3). Non-average behaviour represents new potential in a probability landscape resulting in phenomena similar to the information vortices noted above. The size of the probability hill (or information node) is dependent less on the initiator of a new behaviour than the manner in which it is perceived by others. This perception is a compilation of reflexive decision making (Giddens 1991), the position of the individual in the social structure (Archer 2000), the history of past innovations (Allen 1994; Allen and McGlade 1987) and the stability of the system. If the system is close to criticality (i.e., unstable), small internal perturbations (i.e., non-average behaviour) will be perceived much differently from in stable systems and will have a bigger impact on the system as those participating in average behaviour begin to align themselves with the innovators resulting in positive feedback and growth in the size of an information vortex. Thus the size of the potential hill (or information node) on the probability landscape will be considerably different for the same perturbation in different systems, depending on the closeness to criticality. This process can be seen in community dynamics and quasi-group formation in the ethnographic record.

Factionalism and Quasi-Group Formation at the Hopi Village of Orayvi

The large number of abandoned prehistoric sites and regions throughout the world attest to the fluid nature of both settlement location and community composition, as does the formation of new 'daughter' sites through the 'budding off' (or fissioning) process. Additionally, the variable length that aggregated sites were occupied within the same region in the same chronological period indicates that some communities were more successful at negotiating the tensions and factional tendencies inherent in quasi-group formation than others (Stone 1999). Why some of these communities were abandoned fairly quickly while others were occupied for extended periods has been the centre of debate for some time. An examination of factional activity in ethnographic pueblo communities in the American Southwest within the context of nonlinear modelling may do much to aid in our understanding of this phenomenon.

During historic times, factional disputes within pueblo communities in the American Southwest have been both numerous and well documented. In some instances, factional divisions have led to village fissioning (Ellis 1979; Garcia-Mason 1979; Parsons 1928) and in others, the villages remained united despite factional disputes by experimenting with new organizational structures (Arnon and Hill 1979; Brandt 1994; Dozier 1966; Edelman 1979; Fenton 1957). For our purposes, the fissioning of the Hopi village of Orayvi in 1906 is particularly enlightening (Cameron 1999; Dockstader 1979; Levy 1992; Rushforth and Upham 1992; Titiev 1944; Whiteley 1988). The factional disputes of the village are extremely well recorded, though the reasons for the rise of the factional dispute, the reasons why followers chose the side they chose during the dispute and the reasons for the ultimate fissioning of the village are highly debated. Additionally, other Hopi villages in close proximity to Orayvi that were undergoing similar external pressures did not fission at the same time, making Orayvi a particularly good case study for examining the interplay of internal and external factors and the impact of initial conditions in the process of village fissioning. A multi-causal, nonlinear approach that pays particular attention to initial conditions, system criticality, and the nature of information vortices can help us to understand this process better.

Prior to the fissioning of Orayvi in 1906, it was one of six villages located along the southern edge of Black Mesa (Figure 5.1), and the only village located on Third Mesa. The other villages were Polacca, Walpi, Haano, Musangnuvi, and Songoopavi. After 1906 (and continuing today), there were four villages on Third Mesa and (not counting remote farming communities and agency towns) 11 Hopi villages (Figure 5.1). The actual events of the 1800s and early 1900s that led up to the fissioning of Orayvi are generally agreed upon and well documented by Levy (1992), Titiev (1994), Whiteley (1988), Dockstader (1979) and Rushforth and Upham (1992). The following is a synthesis of the data presented in these sources.

There were two core groups of individuals that formed the heart of the quasi-groups at Orayvi. Based on interaction with the US government, they were labeled as 'Friendlies' (also known as Progressives) and 'Hostiles' (also known as Traditionals) by both the US and the Hopi themselves. In reality, neither group actively sought interaction or compliance with US government decrees, and appear to have differed mostly in terms of the stalling tactics they used when dealing with government demands for acculturation (Levy 1992). The core individuals of the quasi-groups were identified as early as 1880 when Loololma, head of the Bear clan, became the head of the village, or *kikmongwi* (Dockstader 1979; Levy 1992; Rushforth and Upham 1992; Titiev 1994; Whiteley 1998). His style of interaction differed considerably from his predecessor (his father), in that he took a more non-confrontational approach to dealing with government officials and was thus seen as a 'Friendly'. In reality, core groups of individuals probably formed the centre of loose quasi-groups before this and would become increasingly solidified, eventually forming the factions that split the village in 1906. The leader of the core group that challenged Loololma (i.e., the 'Hostiles') was Lomahongiwma, head of the Spider clan. Levy's (1992) excellent analysis

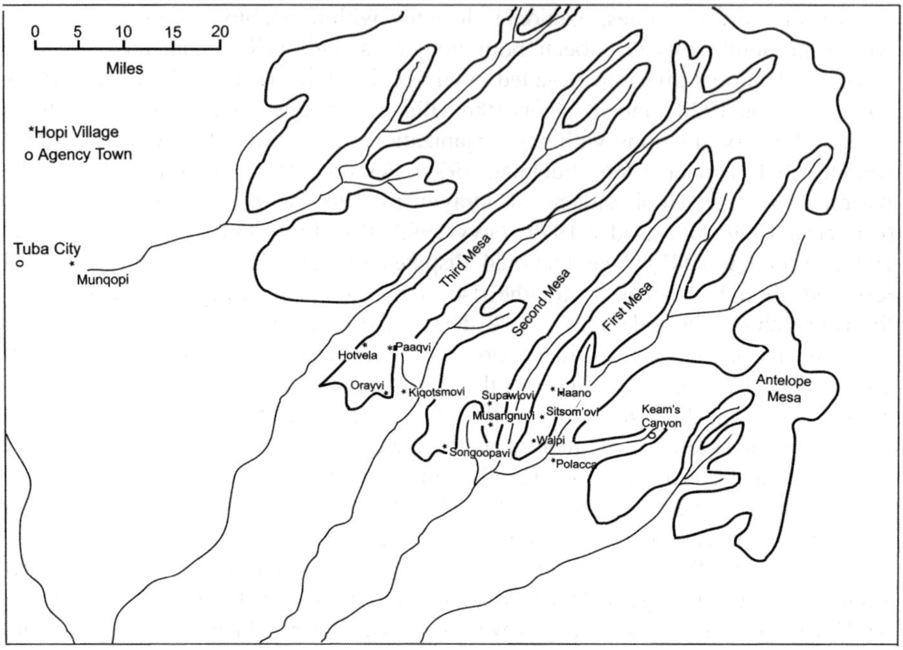

Figure 5.1 Location of Hopi villages

demonstrates that Loololma and his close relatives were able to attract followers who favoured the status quo to their position, particularly those who had preferential use rights to both prime agricultural lands and ceremonial offices. Conversely, followers who aligned themselves with the Lomahongiwma and his close relatives were, almost universally, those that had the least (or in some cases no) access to agricultural land and ceremonial offices. As such, they were individuals who were challenging the status quo. There are two issues that I would like to concentrate on here with respect to the factional dispute. First, why did the core individuals in the two groups take the positions they did? This is particularly interesting because both the Bear and Spider clans had access to some of the best agricultural lands and the most ceremonial offices. That is, Lomahongiwma and his close relatives that formed the core of the quasi-group challenging the status quo, were the establishment – not the dispossessed. Given this, why did they challenge the status quo? Second, why did the followers that he attracted believe that aligning themselves with Lomahongiwma, given his position as an elite, rather than Loololma would increase their chances of attaining their personal goals? I believe the answers to these questions centre on the control and manipulation of information through the ritual system by quasi-groups and the perception of this information flow by members of the community. To understand this, some information on Hopi socio-political structure is needed.

Hopi is a matrilineal society with village endogamy in which women dominate the economic sphere while men dominate the ceremonial/political sphere. In 1906, wage labour was rare and the Hopi still practiced a subsistence economy (Levy 1992; Titiev 1944). Land was, and still is, distributed to families by the clan. However, the land was not distributed equally to all clans or to all families within a clan. Specifically, certain clans controlled most of the best agricultural soils, others most of the moderately productive fields, and still others no fields at all. Within clans, the woman heading the primate lineage was charged not only with keeping the clan ceremonial paraphernalia, but also with the distribution of land controlled by the lineage. Primate lineages received the most land, followed by alternate lineages (headed by her sisters). The marginal lineages (headed by more distant relatives) received the least land. Brothers of the primate women of the clan dominated the ceremonial offices connected with the ceremonies the clan controlled, though members of other clans served as participants of the ceremony as well (Levy 1992). Whereas clan and phratry exogamy distributed the economic and ceremonial control through the village to some degree, it was not equally distributed and some families and individuals had more access to economic and ceremonial positions than others. The Hopi refer to these individuals as *pavanisinom*, which roughly translates to the upper class, while those with little or no land and access to ceremonial office are referred to as *sukavungsinom*, or commoners. This differential distribution of resources is reinforced ritually, through the migration myths. That is, the most powerful clans in terms of land tenure and ceremonial office were the first to arrive at the village and others came later.

In years of sufficient precipitation, when sufficient agricultural products could be produced, there was little apparent friction over the distribution of land. However, in times of drought, lower-ranked clans and lineages would face subsistence shortfalls that could not be overcome by the limited sharing that occurred in the village (Levy 1992). During times of drought (which existed from 1865-1904 and were particularly severe from 1866-1869, 1892-1904 and in 1906 [Table 5.1]), several options were available. The first is the use of fields at greater distance from the pueblo. The numerous field houses and seasonally occupied farming villages that are evident both archaeologically and in historic times are a testament to this. In times of severe regional drought, the entire region might be abandoned. In the late 1800s, regional abandonment was no longer an option due to the imposition of the reservation system. The formation of farming villages in other areas of the reservation could still be undertaken. For example, the farming village of Munqapi was inhabited by marginal lineages from Orayvi in the late 1800s. However, beginning in 1875 and continuing until the 1906 split, this option was also hindered due to the increasing encroachment of Navajo and Mormon settlers on Hopi land. Added to this was a series of smallpox epidemics in 1853, 1866, and 1898 and increasing pressure from the US government for Hopi children to attend US sponsored day schools.

Table 5.1 Timeline of events at Orayvi

1851-1865	Above average rainfall.
1853	Smallpox epidemics throughout Hopiland.
1865-1904	Below average rainfall.
1866	Smallpox epidemics throughout Hopiland.
1866-1868/1869	Severe drought.
1875-1906	Navajo incursions into Hopiland, particularly at outlying farming villages and springs.
1887	Dawes Act Passed.
1887	Boarding School for Hopi children established at Keam's Canyon.
1888-1903	Mormon incursions into Hopiland, particularly at outlying farming villages and springs.
1890	Quotas for school attendance at Keam's Canyon set by US government.
1891	U.S. Surveyors' map land for allotment under the Dawes act. Orayvi men pull up the stakes at night. US government arrests Hostile leaders for the action.
1892	Government day school founded at Orayvi.
1892-1904	Severe drought.
1894	Land allotments under the Dawes act abandoned for Hopi.
1894	Dispute between factions at outlying farming communities leads to incarceration of Hostile leaders at Alcatraz for one year.
1898	Smallpox epidemics throughout Hopiland, particularly First and Second Mesa.
1905	Drought broken with rains.
1906	Severe drought – Orayvi's main spring fails.
1906	Orayvi fissions.

Source: J. E. Levy, *Orayvi Revisited, Social Stratification in an 'Egalitarian' Society*, (1992); S. Rushforth and S. Upham, *A Hopi Social History* (1992); and M. Titiev, Old Oraibi, a Study of the Hopi Indians of Third Mesa (1944).

In other words, several major sources of stress existed in the larger environment at the time of the factional dispute. These environmental stresses alone are not sufficient to explain the political conflict in Orayvi, however, since these stresses were present at the same, or in some cases greater (particularly with regard to the smallpox epidemics), levels at the other Hopi villages near Orayvi (Figure 5.1). Yet none of these villages split or had the well-solidified factions evident at Orayvi at this time. Therefore, we must look internally, to the social and political interaction between the inhabitants in the village to understand this process, how the inhabitants viewed the initial conditions, and how this perception served to dampen feedback loops at the other villages but heighten them at Orayvi.

Loololma's position as the political leader of the quasi-group and later solidified faction that supported the status quo is understandable given his position as *kikmongwi* and his sisters' positions of economic dominance. Lomahongiwma's position as leader of the quasi-group and then solidified faction that challenged the status quo is more problematic on the surface, though deeper examination resolves these issues. Lomahongiwma was head of the Spider clan, 'the highest-ranking clan next to, or along side of Bear clan' (Levy 1992:137); an officer in the Antelope, Blue Flute and Momtsit ceremonies (more ceremonies than Loololma); and a phratry brother of Loololma. As a powerful member of the *pavansinom*, why would he become a core individual challenging the status quo and why would the *sukavungsinom* follow him? The answer to this question lies in differing personal goals of individuals within Hopi society, which in turn were based on their place within the structure of that society and differential access to information. In other words, not only do personal goals vary from person to person, but so does the type of agential action an individual can reasonably take, due to their place in the power structure of the society (Archer 2000). Core individuals in quasi-groups that successfully challenge the status quo are likely to be individuals who have power in the existing system because they are in a good position to successfully manipulate the system (Brumfiel 1992), particularly the ritual system and symbols within it. Additionally, they tend to have an existing power base to build upon to challenge those with more power due to their access to information systems connected with ritual (Crick 1982; Lucero 2003). Finally, while Lomahongiwma was a *pavansinom*, he was not the *pavansinom* with the most power, and he challenged the one who did have the most power, the *kikmongwi*, Loololma.

The conflict between the core individuals of the factions was played out largely in the ceremonial realm. As stated above, ritual is a particularly powerful medium for core individuals to express their positions, attract followers, and solidify quasi-groups into political factions because of its role in information processing. Additionally, it is a medium for followers to express support for one quasi-group or another. Aside from the socially oriented and performance nature of ritual noted above, there is the fact that ritual symbols are 'condensed'; when used in ritual situations, symbols explode and carry with them references forward, backward, and convey social orders sanctioned by the supernatural (Rothenbuhler 1998). In the case of Orayvi, the explosion of condensed symbols is particularly noticeable. The first aspect of ritual that entered the dispute occurred in the 1880s, before the quasi-groups solidified into the factions that split the village in 1906. This came in the form of variations in the creation and origin myths that linked the various clans (in particular Bear and Spider) to the actions of supernatural figures (Levy 1992), and thus legitimized their primate positions while delegitimizing others. These variations in myth were repeated in rituals controlled by the different quasi-groups and participated in by the village as a whole.

The control of ceremonies by the different quasi-groups involved, and the repetition of the variations in myths on these socially sanctioned stages, plus other information processed through the condensed symbols of ritual, created an information vortex on the social landscape of the community. The centripetal forces tied to information vortices can be seen operating at Orayvi, from the 1880s on. Specifically, ritual dissonance expanded and became increasingly formalized, leading to a shift from fluid quasi-groups to defined factions. By 1891, both Loololma and Lomahongyoma claimed to be the leader of the Soyal ceremony, the ceremony that both set the ceremonial calendar for the year and reaffirmed the position of the *kikmongwi* (Rushforth and Upham 1992). From 1891 on, two Soyal ceremonies were held, resulting in two different ceremonial calendars being set. Concomitant with this, duplicate ceremonies were held throughout the year (Dockstader 1979; Levy 1992; Rushforth and Upham 1992; Titiev 1944). Followers were forced to permanently align themselves with one core group of individuals or another or risk being left out of the ritual cycle and the information processed during the rituals. As a result, the quasi-groups solidified into well defined factions. As this solidification occurred, the different factions not only held separate ceremonies but interfered with the ability of the other to hold theirs by denying access to ritual structures and paraphernalia, causing both sides to find alternatives not stated in, but justifiable through, the creation and origin mythology. In the economic sphere, this solidification of factions was seen in challenges to the traditional land tenure system. Specifically, each group planted crops on lands claimed by the other through the traditional land tenure system. This was further exacerbated when the Hostiles invited a group of clansmen sympathetic to their position from the village of Songoopavi on Second Mesa and told them to farm land claimed by individuals in the Friendly faction (Levy 1992; Rushforth and Upham 1992).

The final split of the village (resulting in all of the individuals in the hostile faction leaving) occurred in September 1906, and was, again, played out symbolically, in a ritual pushing match. The men of the Friendly faction went to the home of the leader of the Hostile faction and ordered the Songoopavi residents to leave the village. When they refused and the Hostile faction supported their Songoopavi allies in their refusal, a 'scuffle' broke out among some of the individuals present (Titiev 1944). By late afternoon, the men of both groups faced each other on the north end of the village. The leader of the Hostile faction drew a line in the dirt and stated that if the Friendly faction could push them across that line, they would leave peaceably. But, if the Hostile faction pushed the Friendly faction across, they, and their Songoopavi allies, would stay (Levy 1992; Rushforth and Upham 1992; Titiev 1944). A shoving match ensued and the Hostile faction was pushed across the line. By that evening, all men, women and children associated with the Hostile faction left the village and travelled north on Third Mesa to establish a new village.

The question remains: how did the quasi-groups of the 1880s solidify into the factions of Orayvi that led to the eventual split when this did not occur at other Hopi villages undergoing the same external stresses? The answer to this is context specific and lies in differing initial conditions. Specifically, I argue that one of the major factors was the perception of the followers regarding the stability of the system. If stability is defined in the tradition of homeostatic models that concentrate only on the environment, the other villages would appear to be more unstable and susceptible to factional disputes than Orayvi. In addition to the prolonged drought at these other villages, the demographic impacts of the smallpox epidemics of the late 1800s were greater on First and Second Mesa than on Third Mesa (Levy 1992) because of their closer proximity to agency towns. This proximity also led to greater acculturation pressures, particularly regarding forced compliance with government set quotas for school attendance, which was higher among First and Second Mesa children. However, if we use the definition of perceived stability given above (i.e., the ability of individuals to gain access to the information needed to achieve their personal goals), a very different picture emerges. Because smallpox does not discriminate against high- and low-ranked clans and lineages, access to information, and thus political power, through ceremonial office was more fluid as some clans died out and others increased in size in the villages on First and Second Mesas. Therefore, while the same acculturative and ecological stresses were evident, the initial conditions were *perceived* very differently in the villages, resulting in a state of criticality (as defined by Bentley and Maschner 2001) at Orayvi but not at the other villages. Given this state of criticality, any actions taken by the core members of the quasi-groups would be perceived and reacted to in a deviation enhancing (rather than a deviation dampening) manner by the followers at Orayvi. The combined affects of environmental stress, competition between the core members of quasi-groups within the community for power, and perception of the followers resulted in a system dynamic that is greater than the individual parts – i.e., a chaotic system – that reached a bifurcation point and fissioned.

Lessons for Archaeology

How does this very interesting case study from the ethnohistoric record help us to better understand factional formation in prehistoric communities and how can we apply, in a practical manner, the lessons we learned from Orayvi? This is a particularly compelling question since much of what we can dissect about the Orayvi case in terms of the movement from quasi-group formation to solidified political factions that split the village apart is related to the actions of individuals at particular instances. Given the coarse resolution of our chronological data and our inability to identify transitory individual actions from the prehistoric record, how can we begin to dissect a prehistoric case in a similar manner? This question can be answered at both the general, paradigmatic level, as well as addressing more specific questions of village fissioning and community dynamics in middle range communities.

At the general level, Orayvi is an example of the phenomenon of quasi-group formation and the nature of information vortices. Further, Orayvi demonstrates the impact of individual and group agential action constrained by societal structure and the importance of context specific initial conditions. Additionally, while most of the data about the Orayvi split is at a level of detail that is difficult to achieve from the archaeological record, the aggregate result of those individual events is evident over a long period of time (more than 25 years prior to the final split at Orayvi) and through the entire community. However, we cannot begin to understand this type of aggregate behaviour and how groups reach bifurcation points unless our theoretical and methodological approaches appreciate the importance of integrating multiple scales of analysis and the interplay of individuals and groups within societal structure (cf. Spencer-Wood 2000). Additionally, culture cannot be broken down into subsystems that are examined separately if we are to understand culture change. Rather, we must look at the interplay of different scales and different aspects (economic, ritual, political) of the culture simultaneously within an historical context because the whole is more than the sum of the individual parts (cf. Kus 1981). Finally, Oravyi demonstrates the importance of looking internally as well as externally to understand system dynamics rather than just externally driven homeostatic models (cf. Brumfiel 1992; Stone 1999).

At a more specific and practical level, the Orayvi case is instructive of the process of village fissioning in middle-range societies. When Orayvi fissioned, almost half the population left to form a new community. The formation of new communities in this manner is extremely common in prehistory. It is not sufficient to describe this phenomenon as daughter villages budding off from parent communities. We need to examine the *process* of village fissioning if we want to understand community dynamics.

Scholars in physics, economics and ecology (Allen 1982, 1989, 1994, 2001; Clark, et al. 1995) argue for moving away from Newtonian, mechanical models that concentrate on 'normative', or average, behaviour and its necessary emphasis on individual/agent similarity if we are to understand the internal dynamics of systems and how systems change. Instead, they argue we need to concentrate on variability and the integration of multiple scales of analysis (see also Spencer-Wood 2000). The idea of information vortices played out through competing ritual structures is a particularly useful way of examining this process and eminently doable through the examination of ritual architecture. Specifically, we need to examine variability in ritual architecture and paraphernalia both within and between communities (Stone 1999, 2002) to better understand community dynamics. By concentrating on variability and viewing communities in a dynamic manner we can begin to see cycles of increased quasi-group action in prehistory and gain an appreciation of the complex nature of socio-political interaction in middle-range societies.

Acknowledgements – This chapter represents issues I have thought about and dealt with tangentially for some time. The impetus to finally pull together the theoretical issues involved in factionalism and nonlinear modelling was a symposium I participated in titled 'Nonlinear Systems Approaches in Archaeology' organized by Christopher Beekman and William Baden at the 2002 AAA meetings. I would like to thank them for organizing the symposium which prompted me to write this chapter. Additionally, I am much indebted to Susan Dublin, who read and gave valuable comments on an earlier draft of this chapter. As always, however, all shortcomings remain the responsibility of the author.

Chapter 6

Modelling Prehistoric Maize Agriculture as a Dissipative Process

William W. Baden
Indiana University – Purdue University Fort Wayne

Maize is a very exhausting crop; scarce any thing exhausts the land more.
(Mitchell and Young 1775:52)

Introduction

The research framework associated with complex adaptive systems (CAS) differs from more traditional, reductionist approaches in a number of ways. CAS concepts of emergence, fitness landscapes, self-organization, non-equilibrium conditions, etc. change our research objects and their observational scope. Single, linear cause and effect relationships no longer have relevance when our goal is an understanding of emergence. Even earlier systems theory approaches, after describing societies as composites of interconnected components, tended to regress (no pun intended) to simple dependent-independent analyses. Understanding emergence requires a simultaneous evaluation of cumulative individual actions (agents) operating within a maze of boundary conditions (limitations/costs). This convergence of behaviours and situations is interpreted as a fitness landscape (Kauffman 1993:39) which simultaneously serves as the evolutionary specification of organisms and their environments. A landscape becomes a metaphor for an organism's (or society's) evolutionary trajectory over time when it is quantified as a cost function visualized across a three dimensional surface of hills (high costs) and valleys (low costs). 'Fit' entities will seek out and take the path of minimal, local cost. Translating this search process into an algorithm facilitates the use of computer simulations – the principle operationalizing tool of CAS.

This discussion focuses on specifying a major component of fitness landscapes for prehistoric agricultural societies. I will examine the maize-based populations of North America's Eastern Woodlands. As an energy extraction process, agriculture can be modelled as a dissipative system within a non-equilibrium thermodynamic framework. The fitness potential of maize growers is directly linked to their ability to sustain an adequate yield within a dissipative environment. Their agronomic choices and the environment's response will determine the sustainability of their societies. A measure of stability can be expressed by their potential to withstand fluctuations in yield or more exactly, reductions in potential yield. Unstable conditions provide an opportunity for societal adjustments and the realization of emergent phenomena.

Applying thermodynamic principles to anthropological research is not new. Allen (1997b) has noted that the second law appeals to anthropologists (and others) with its acknowledgement of 'time's arrow' and a process of change. Its use also taps into a latent scientific legitimacy that tends to seduce social scientists. Allen felt that our applications of the second law were inappropriate in that we missed the uniquely creative nature of cultural evolution. In other words, societies are not mechanical and they certainly are not closed systems – the necessary precondition for second law applications. I would add to this critique the failure to incorporate the more applicable non-equilibrium realm of thermodynamics and its characteristic dissipative structures.

Earliest Abuses of the Second Law

Many anthropologists have focused their theoretical perspectives on energy flow within societies using heuristic devices like classical thermodynamics, catastrophe theory, deterministic chaos, etc. Working within an energy extraction framework has produced many useful insights even though most of these exercises have primarily invoked the thermodynamic laws metaphorically. In fairness, the deceptively simplistic nature of the first two laws can mislead us into misapplying them to social phenomena:

1. In an isolated, irreversible system energy is conserved,
2. In an isolated system disorder/entropy is always increasing.

The temptation to invoke these laws is not surprising given the often cited importance of rationalizing the impact of the Second Law's object – entropy – on biological and cultural evolutionary processes. I want to re-examine the usefulness of applying thermodynamic concepts to cultural contexts in light of more recent understandings of nonlinear systems. The end result may still retain a metaphoric feel but hopefully it will also extend the earlier studies by broadening our research strategies and creating a firmer foundation for self-organizational perspectives.

The earliest associations between social order and thermodynamic principles originated in the Darwinian era of the 1860s. The Victorian convergence of 'survival of the fittest' and energy theories was deeply entwined with 19th century social thought. Interestingly, the emerging field of physics used social metaphors to educate the populace on the relevance of physical theories in numerous scientific popularizations (Myers 1989). The concepts of energy and entropy were seen as intuitive. After all, mechanical systems clearly require energy to work and over time run down. Conversely, social writers of the time, like Henry Adams, used thermodynamic concepts to add exactness to their descriptions of social degradation: societies, like steam engines, will eventually stop and their restart will require external forces/energy. Perhaps the most prolific authors were those arguing a 'natural theology' (the precursor of today's 'creation science') (Wilson 1989). Promoters, like William Thompson (Lord Kelvin), used the implications of energy dissipation to support their view of a Divine hand in the creation and maintenance of the universe. The tone of their arguments was similar to:

> As in the social world a man may degrade his energy, so in the physical world energy may be degraded; in both worlds, when degradation is once accomplished, a complete recovery would appear to be impossible, unless energy of a superior form be communicated from without. (Stewart and Lockyer 1868:322)

The point is, use of thermodynamic metaphors is inherent in the development of both physical and social sciences. It should not be surprising that continued use inspired numerous anthropologists of the 20th century.

Modern Thermodynamic Precedents

Although certainly not the first to associate cultural development with energy extraction (see White 1954), Leslie White was anthropology's most significant 'prophet' of the Second Law (1943, 1949, 1959). In advancing his theory of cultural evolution White noted the trend of *per capita* energy consumption along a complexity continuum. The more complex a society the more energy distributed per member. White concluded that the process of cultural evolution was really all about changes in the efficiencies of energy extraction by succeeding societies. He equated (1959:47) the relationship with a hand-waving (i.e., non-rigorous) formula:

$$\textbf{E}\text{nergy X }\textbf{T}\text{echnology} \rightarrow \textbf{P}\text{roduct} \qquad (6.1)$$

In which energy (**E**) could be divided between human and nonhuman components.

Entropy for White was the measure of the unavailable energy in a [irreversible] thermodynamic system (1959:33n). Following R.J.E. Clausius, entropy in such systems is constantly increasing. Intuitively, for White, human societies run counter to the universe's trek toward entropic randomness. For societies, order is not only the rule but increasing the complexity of that order over time is a common characteristic of cultural evolution. White had discovered the paradox of applying the Second Law to social systems – on face value it does not fit.

Undaunted, White cited Ludwig Boltzmann's observations that life is the process of obtaining and incorporating free or available energy from the environment (White 1959:34). To counter the paradox, White was attracted to Erwin Schrödinger's conceptualization of this free energy as *negative entropy*. By extracting more negative entropy than they lose (i.e., positive entropy), societies seem to defy, in a sense, the Second Law. In Schrödinger's words, living organisms suck 'orderliness out of the universe' (1944:71-72). Thus, White was able to move past the paradox and concentrate on the factors in his formula, including the addition of an environmental variable (1959:49):

$$\textbf{E}\text{nergy X }\textbf{T}\text{echnology X en}\textbf{V}\text{ironment} \rightarrow \textbf{P}\text{roduct} \qquad (6.2)$$

Application of this theory clearly involved focusing on those cultural traits related to the components of energy extraction processes. For any given culture, **E** represents subsistence options; **T** corresponds to tool sets; and **V** encompasses an environmental context. In reality White's exercise became more classificatory than explanatory – a problem noted by critics like Julian Steward (1960) – with the application of an energy twist to the 'vulgar' Tylor-Morgan approaches.

Despite its inherent weaknesses and negative undertones, White's approach still seemed intuitively useful. Retooling White's evolutionary theory into an observation rather than a law – levels of social integration are proportional to thermodynamic achievements (Sahlins and Service 1960:36) – minimized the negative classificatory undertones. But the fact remained that significant cultural transformations are correlated with energy capture transitions. The observation's practicality was promoted by White's chief 'disciple,' Betty Meggers (1954, 1960), with a particular focus on environmental reconstruction and agriculture. She would argue that limits to agricultural potential largely define the achievable level of cultural development for any advanced society in a specific context. Environmental degradation as a result of soil erosion and climatic fluctuations would provide the quantification of **V** (Eq. 6.2) and a set of observations worthy of anthropological examination.

Another interesting, energy inspired broadening of research strategies was offered by Shawcross' subsistence analysis under the rubric of the First Law – the conservation of energy (1972). He believed that society's transformation of energy into other forms qualified as First Law phenomena. His primary criticism of earlier thermodynamic approaches which emphasized the Second Law was that they were not quantifiable. This was a reasonable criticism given that entropy – a system's unusable energy – is not easily observed. Shawcross reasoned that the archaeological record was the end product of an energy-matter transformation process restricted by measurable production efficiencies. If we could translate site content (subsistence evidence) into energy measures, limiting their realizable potential based on extraction efficiencies, we should be able to estimate population parameters, like carrying capacities. He demonstrated such an application using shellfish remains from New Zealand Maori sites.

Of all the past theoretical excursions into the relevance of thermodynamic perspectives perhaps the most balanced was that of Richard Newbold Adams (1981, 1982a, 1982b, 1988). Adams recognized that conflicting views on the applicability of thermodynamic principles existed within the physical sciences and anthropology's uses had been largely metaphoric (1988:xvi). But he objected to discarding the approach arguing that thermodynamic concepts may help us understand social dynamics. Just as in the 19th century, if the metaphors help us expand our research strategies and place new sets of observations and relationships within our view, their construction is worthwhile.

Adams argued that we should be identifying *dissipative structures*. What are dissipative structures? Based on the theoretical research of Ilya Prigogine and others (1961, 1962, 1971; Glansdorff and Prigogine 1971; Nicolis and Prigogine 1977), a dissipative structure is a system existing as ordered configurations which emerge beyond instabilities in the thermodynamic branch (Nicolis and Prigogine 1977:60). That is, systems (e.g., chemical reactions, ant colonies, societies, whatever) existing far, far from thermodynamic equilibrium in a temporal-spatial context where nonlinear feedback relations apply. These systems display unique, nonlinear self-organizational characteristics in reaction to instabilities (Glansdorf and Prigogine 1971:73). More importantly they expend (i.e., dissipate) free energy. Adams used dissipative structure arguments to map and explain changes in Victorian England (1975, 1982a, 1982b, 1988).

Any theory, especially if its intent is to define a 'law' that is more observational than predictive or explanatory is doomed to be underutilized and eventually replaced. This was the fate of White and his followers' attempts to invoke the Second Law. It was not an easy sell to argue societies were like gas molecules in a closed system. Their most critical technical mistake was failing to account for the 'isolated system' restraint on the applicability of the three laws. Cultures, after all, are very open systems. Yet the effort did leave new research strategies in its wake. Today's faunal, botanical, and environmental specialists owe some portion of their professional existence to the energy debates of the 1950s and 60s.

Later discussions of dissipative structures recognized a new set of energy observations and relationships that more correctly dealt with the open nature of cultures. Dissipative structures, by definition, exist in non-equilibrium environments. However, without the demonstrated ability to clearly measure constituent components, like 'entropy,' or to define 'non-equilibrium' in cultural terms they have had limited applicability, too. We appear to still be relegated to using metaphors to describe phenomena that intuitively seem to follow thermodynamic principles. We need to reincorporate many of these intuitive relationships into a more holistic approach that is truer to the thermodynamic paradigm. To demonstrate such an approach I will detail the nature of dissipative structures as they apply to maize agriculture practiced by prehistoric Mississippian cultures of eastern North America.

Dissipative Structures

Recall from White's legacy that it is all about capturing free energy. But this energy, what Schrödinger thought of as negative entropy, comes from outside the system. For Mississippian societies, agriculture along with hunting, gathering, and firewood procurement combine to provide their *external* source of negative entropy (which ultimately can be traced back to the Sun). Through their processing of this energy, members of Mississippian societies maintained themselves as individuals and in so doing ensured social reproduction of their culture. Surplus free energy that can be converted into new matter, like mounds and tools, is contained or slowly dissipated in the form of culturally defined order and not expelled immediately as waste; hence the reason it can be considered negative in the overall balance equation:

$$dS = d_eS + d_iS \tag{6.3}$$

where the total change in entropy, dS, is the sum of changes in the external flow of entropy, d_eS, into the system and the internally produced entropy, d_iS, in the form of dissipated waste.

We know from the Second Law that $d_iS \geq 0$ and for an isolated system $d_eS = 0$. But for an open system the external component, d_eS, lacks sign restraints and can therefore be negative. Such a system maintains a steady state whenever $dS = 0$. So, sufficient negative entropy flow can maintain order in a system if and only if the system is *not* in equilibrium. Why? An equilibrium state necessarily implies $d_iS=0$ which means (Eq. 6.3) also equals 0; hence d_eS would have to equal 0. For a thriving open system this would be impossible (recall that White noted a system where $d_iS=0$ is one that cannot do work, i.e., it has reached thermal equilibrium – a unique steady state condition sometimes referred to as death). The fact that both entropy components are non-zero implies that open systems, existing far from equilibrium ($d_iS >> 0$), have the thermodynamic capability to *create order*, i.e., produce an overall $dS < 0$. But a system does so by also creating instabilities in those subcomponents that respond within the near equilibrium thermodynamic realm by increasing disorder over time (remember, $d_iS > 0$ for a non-isolated society) (Nicolis and Prigogine 1977:19-25).

To conceptualize cultures under the dissipative structure paradigm we must first isolate what is internal. The internal components consist of the people and the culturally defined socio-political constructs which define a society. Internal elements lend themselves to numerous analytical aspects of today's complexity sciences, like agent based modelling. We next separate the external inputs such as subsistence, energy, and raw material resources – the building blocks of material culture. The interface between the internal and external defines the *boundary conditions* of the system. Recognizing that every culture exists precariously far from thermodynamic equilibrium ($dS \approx 0$ when $d_eS << 0$ and $d_iS >> 0$), our analyses should focus on these boundary conditions. It is at this interface that changes in the flux of incoming energy *may* force self-organizing adjustments in the internal social-political order in an effort to maintain the culture's total entropy balance – in simpler CAS terms, the point of emergence. Significant structural changes will appear as a cultural phase transition. In extreme situations the result can be total structural collapse, hopefully to a lower entropy level that better matches current d_eS input (i.e., d_iS decreases). If we believe in the cultural relevance of dissipative structures and their self-organizing capabilities then we must concentrate on identifying fluctuations in the boundary conditions.

Applying the above concepts to the analysis of prehistoric societies requires, in the obvious absence of empirical observations, the use of simulations. Before we can accurately simulate culturally defined dissipative structures we need to be able to model the external structures that supply the energy surpluses. These structures are also dissipative in nature, though external to their cultural counterparts. I will argue that if we look at agriculture from the perspective of the parent culture, the subsistence components (i.e., the field and crop resources) lay outside the cultural realm as defined by thermodynamic principles. Granted they more typically are seen as integrated components of the socionatural system, but in a thermodynamic context this integration is actually part of the boundary conditions that join the two realms. In this sense they are perhaps more accurately referred to as *directed* dissipative structures with the cultural realm guiding the natural systems. This also implies the existence of dissipative structures inside the agricultural processes. For the math to work, we must recognize this distinction.

Finally, consider the conditional '*may* force self-organizing adjustments' outcome. We are really looking at emerging cultural complexity when we examine dissipative structures and the creation of order in response to fluctuations far from equilibrium. Sufficiently complex systems can be assumed to be in a metastable state, especially when far from equilibrium (Nicolis and Prigogine 1977:462-463). But no society is truly stable given the flux of information, energy, populations, etc. they are continually exposed to. Still a fluctuation that can cause a shift to an alternate stable state would need to be significant. Students of nonlinear systems represent objects of study and their collection of observed attributes as *phase space* – a Cartesian-like representation of objects (e.g., societies) mapped to coordinates along attribute dimensions. In a truly deterministic world these objects would be represented by points but, as alluded to in Chapter 1, our inability to precisely quantify each of these dimensions results in more of a fuzzy volume representation than a point. At time t=0 this volume defines the initial state of the object or collection of objects that 'appear' to be alike and its size correlates with the precision of our observations. The archaeological concept of Mississippian or any single phase within the tradition is an example of such a volume. Statistical mechanics refers to such collections as statistical ensembles (Nicolis 1995:58) whose trajectories take the form of discrete states rather than a point's continuous path. It is within these bounded regions of phase space that fluctuations occur differentially to each member object's phase space coordinates. Through our statistical viewfinder they seem to leap from one location (i.e., state) to another with a predictability constrained to probability statements rather than well behaved, deterministic equations. If the sequence of leaps appears irregular we refer to it as chaotic.

As we implement these concepts of initial states, apparent randomness, and fluctuations into our modelling we recognize two stochastic probabilities: 1) the likelihood that a fluctuation can occur and 2) the probability that it will result in a significant change in the macroscopic system variables (i.e., nontrivial change). For archaeologists, any discussion of past processes must, therefore, recognize that an observed state or phase transition is one of many possible outcomes. Although some specific self-organized components may be deterministic in nature, generally succeeding outcomes will be a result of restricted choices that appear to us as coin tosses. These repeated bifurcation events combine to trace a society's path, like footprints, through their fitness landscape. When designing models and simulations we must incorporate fluctuations as random-looking events and describe our results in terms of phase transition probabilities. This is especially important in the presentation of simulation results. One execution does not constitute a reliable approximation of the *probable*, real-world result. As a reflection of our precision, simulations should recreate the collective behaviours of the volume and not points. This is achieved using repetitive results to estimate the probabilities.

This discussion is important not only in specifying how models should be constructed but it also provides a clarification of the concept of chaotic systems. The use of the term, chaos, is more a reflection of our observational inabilities than of random behaviour. Interpreting a society's state transitions as chaotic does not imply individual behaviours are random even if they necessarily must be simulated by stochastic programming algorithms.

Returning to the thermodynamic aspects, for most Mississippian societies the source of surplus energy was maize agriculture. Maize represented a source of free energy which was harvested, consumed, stored, and converted into new matter. The new matter included hoes, structures, elaborate trade items, platform mounds, palisaded villages, etc. In an effort to operationalize dissipative structures I will examine maize agriculture as a thermodynamically defined system composed of dissipative structures. New levels of order within the agricultural system *may* require, from the parent culture's perspective, social-political responses. To anticipate the impact of this relationship on Mississippian self-organization, we should first address Mississippian choices.

Mississippian Behavioural Choices

Mississippian societies were far too diverse to be reduced to simple generalizations. But there are two important characteristics that most shared:

1. Most regions reveal three major phases: Emergent (ca. A.D. 750-900 to 1000), Middle (ca. A.D. 1000 to 1300), and Late (ca. A.D. 1300 to 1600) (actual regional dates will vary).
2. Most regions experienced an apparent collapse or widespread site abandonment at the end of their Late phase.

As perhaps the most complex prehistoric manifestation in the Eastern Woodlands, Mississippian remains have been studied for over a century with particular emphasis on explaining their origins and disappearance. Their development can be viewed within the thermodynamic perspectives outlined above: as the growth and reduction of order over time. My focus on the contribution of agricultural processes as a trigger for the observed changes in Mississippian society is not new. Attempts to explain the eventual collapse of such complex societies have occasionally involved agriculture and soil degradation's possible contribution. For example:

> The reasons for the abandonment of hamlets were probably varied but may have centred on both the depletion of natural food resources and on soil fatigue by unrestricted crop-growing. (Harn 1978)

Likewise, numerous ethnohistoric accounts (e.g., Bartram and Van Doren 1928:315; Lafitau 1977:69-70; Sagard 1939:92-93) note the impact of maize farming on Native American settlement systems. Francois du Peron, ca. 1639, writes of the Huron:

> The land, as they do not cultivate it, produces for only ten or twelve years at most; and when the ten years have expired, they are obliged to remove their village to another place. (Thwaites 1896-1901:15, 153)

What behaviours are relevant to recreating such a unique prehistoric system through simulation? Previous discussions (Baden 1987, 1995, 2002; Baden and Beekman 2001) and responses (Foster 2003; Muller 1997:255-257) detail a model building process that looks at ethnohistoric accounts to define the available choices. The resulting model of prehistoric maize agriculture will then need to be evaluated by modern agronomic premises.

The primary sources include the accounts of the French Jesuits and Recollects in New France during the seventeenth and eighteenth centuries. Of the nearly 400 references to maize in Thwaites' (1896-1901) volumes, the relations of Le June (ca. 1635-6), du Peron (ca. 1639), and Rale (ca. 1723) were found to be the most useful. The more detailed accounts of Sagard (1939), Lafitau (1977), Le Clercq (1968), and Lescarbot (1968) also provide input. The writings of Bartram (1853; Bartram and Van Doren 1928) and Adair (1930) serve as major sources for the Southeast.

First hand observations of Native American agriculture are seldom lengthy, even when made by a botanist like Bartram. The first-hand accounts benefit from secondary sources devoted more exclusively to summarizing subsistence activities. The data summaries of Will and Hyde (1917), Herndon (1967), Parker (1968), and Holder (1970) serve such a purpose. When combined, the historical record provides a pattern of early agricultural practices sufficiently detailed for our purposes. In addition, the observations are supported by the specific examinations of others (Baker 1974; Ceci 1975; Day 1953; Heidenreich 1971; Minnis 1985; Rutman 1967).

Information on field size, plant densities, yield potential, and cultivation practices are the most relevant for building a reasonable simulation of farming behaviours. Based on ethnohistorical observations the Eastern North American maize tradition involved the following generalized practices:

1. Fields were cleared using fire one or more years in advance of the first planting.
2. Fire was also used to clear old fields prior to planting.
3. Planting was undertaken after the first sufficient thaw.
4. Three to ten kernels were placed in 'hills' spaced two to three feet apart in rows up to six feet apart; Plant densities were 12,000 to 17,000 plants per acre. More seeds and wider dispersion tended to occur on poorer soils or under drier conditions.
5. No recognized soil fertilization procedures were practiced.
6. Cultivation involved two minimal hoeings when the plants were roughly six inches and knee-high, respectively.
7. Harvesting was undertaken in two phases: the first in middle-to-late summer when the kernels were in the milky stage and the last in the fall after the grain had completely ripened.
8. Practical yield estimates ranged between 10 and 20 bu/acre.[1]
9. Field sizes ranged between 0.3 and 1.5 acres/person (0.12-0.6 ha/person).

We can then generalize that field sizes will be roughly one acre per person (0.4 ha/person), that plant densities would be less than 20,000 per acre (49,400 per ha), that yield potentials will be at or under 30 bu/acre (1883.2 kg/ha), and that weeds would be a significant problem.

Along with farming behaviours, population size and rates of consumption need to be specified. Individual diet dependence on maize can be estimated using isotopic assays of skeletal remains (Bender, et al. 1981). It has been reported that a cline between 35 and 72 percent caloric dependence existed for Mississippian populations (Lynott, et al. 1986:61). If maize provides 3600 calories/kg (Minnis 1985:11), for a 2500 calories/day requirement each person would need 6.47 bu/year or 164.8 kg/year to fulfil a 65 percent dependence (that's 2.54 kg/percentage dependence). Demand curves often take the form of a sigmoid curve – slow adoption followed by a rapidly increasing dependence which tapers off to a maximum. Fitting such a curve to observed percentage dependency data produces:

$$D_t = 13.304 * \tan^{-1}(\pi * (0.005 * t - 6.498)) + 53.54 \quad t \in [900,1700] \quad (6.4)$$

This function duplicates a slow rise in dependence both early (A.D. 900 to 1100) and late (A.D. 1500 to 1700) with a sharp increase starting ca. A.D. 1250 at 44.7 percent. This is a reasonable reproduction of the observed skeletal data (Lynott, et al. 1986). Demand, of course, can fluctuate and not follow such a prescribed path. One adaptation to decreasing yields might be the reduction in

[1] In the US grain is sold by the bushel. A bushel of maize, dried to 15 percent moisture content, is defined as 56 lbs shelled, or 72 lbs husked on the cob. Due to the extensive use of bushels in the historical literature, most references to yield will be in bushels per acre. One bushel per acre is equivalent to 62.8 kg/ha.

maize demand and a shift to wild foods – assuming wild resources could adequately fulfil demand. Isotopic analyses suggest this may have occurred during the Late Mississippian period at Cahokia (Yerkes 2005).

Although not a true 'choice,' population size and growth rates are key factors in determining demand and demand ultimately determines the acceptability of yields. Establishing population size requires an estimate of initial, Emergent Mississippian size and a rate of increase over time appropriate for these populations. Both are difficult to calculate. For rate of increase, the Coale-Demeny life table models (Coale, et al. 1983) provide long term analysis of worldwide populations over the last century. Using their population classification scheme and burial population summaries, we could assume an annual rate of increase (r) lying between 0.003 and 0.017 (3 to 17 per 10,000], which agrees with their West Level 1 tables. We could create an exponentially increasing population function, but unfortunately, defining P_0 becomes problematic. How many people does it take to 'start' a Mississippian social structure? We lack this answer and making estimates for testing would add too much speculation to this study. Alternatively we could assume a fixed population size that can be set to the largest, most stable value possible (under the other constraints of the model). This might be interpreted as an estimate of the population's maize-based carrying capacity. I will take this approach for this example, pending better estimates of Mississippian population sizes.

The final choice parameter involves defining the varieties of maize grown during the late prehistoric. We can be fairly certain that the earliest Mississippian populations grew a form of *Maiz de Ocho* adapted to each region along a route from the Southwest to the Northeast (Diehl 2005; Upham, et al. 1987, 1988). Estimating yield potentials from archaeological remains is difficult due to the lack of complete cobs. Using measurements on cob fragments Diehl (2005) supports early Southwestern yields of 300 kg/ha (4.8 bu/acre) increasing to, perhaps, 400-3000 kg/ha (6.4-47.8 bu/acre) by A.D. 1450. It is not clear what plant densities he is assuming, but densities will have an impact on extrapolating cob potentials to actual yields and optimal densities are variety specific (see below). Analysis of preserved Sinagua fields dating to A.D. 1100 in a marginally productive area of northern Arizona suggested hill densities of 720-760 hills per hectare (Berlin, et al. 1977). Even allowing for numerous plants per hill, this was a very low density planting pattern which reinforces the argument that this area was not capable of supporting Boserup's intensification response (Stone and Downum 1999).

My eastern Corn Belt experiments in growing 8-10 row flint varieties derived from the Northeast (Seneca contexts) and Southeast (Cherokee contexts) suggest plant productivities in the range of 0.035-0.04 kg/plant. Using an optimal density of 35,000 plants per hectare (discussed below) actual yields of 1225-1400 kg/ha (19.5-22.3 bu/acre) would seem reasonable. Based on all these observations I have generally assumed early Mississippian yield potentials to be approximately 500 kg/ha (8 bu/acre). Selecting for fewer rows and shorter plants, later races of Northern Flints would have achieved yield potentials near 1884 kg/ha (30 bu/acre). These yields are consistent with early 19th century farming records (Emerson

1878:38-42, 61) (it was not until the mid 19th century that hybridization practices developed more productive varieties like today's Southern Dents).

Turning yield potential into field size requires matching varietal capabilities with dependence. A 35 percent dependence on 500 kg/ha *Maiz de Ocho* would require at least 0.18 ha/person (0.43 acre/person). A 65 percentage dependence on 1884 kg/ha Northern Flints would require at least 0.09 ha/person (0.2 acre/person). As a general rule of thumb, it is unlikely that more than an acre (0.4 ha) of ground would have been planted per person under any circumstances, but more labour intensive practices (after Boserup 1965) could raise this limit. As a conservative estimate of yield potential, the later yield of 30 bu/acre (1884 kg/ha) will be used in the simulation.

Summarizing the cultural parameters of the simulation:

1. Population size will be set to the maximum value the simulated system can support.
2. Demand will be expressed in terms of kilograms of maize needed to fulfil the dependence need as expressed by Eq. 6.4.
3. Field size will fluctuate but not be allowed to exceed 0.4 ha per person.
4. Weeds will have a major impact on reducing yields.
5. Yield potential, though ultimately determined by the maize varieties grown, will not be expected to exceed 30 bu/acre (1884 kg/ha).

But What About...

Juxtaposed to acknowledgements of soil depletion's effects are a number of agronomic 'myths' that need to be addressed. When researching Native American agriculture one learns that the fields were replenished each year by alluvial soils deposited by spring floods; that by planting beans (*Phaseolus vulgaris*) with the maize, nitrogen was fixed into the soil; and that a reasonably short fallow period sufficiently replenishes the soil. The common conclusion is that soil exhaustion was impossible. I would argue these actions and environmental conditions would not mitigate the negative effects of continuous cropping.

The widespread assumption that the use-life of low river terrace soils is replenished by flooding ignores three points. First, waterlogged conditions encourage denitrification processes. Second, deposited silts, although highly tillable under hoe technology, will not necessarily be nitrogen rich. Very little nitrogen would be found in prehistoric alluvium (as opposed to today's nitrate-rich river deposits). Finally, spring floods will tend to occur at planting time which would increase the risk of crop failure. The real advantage of the alluvial terraces is their proximity to the water table and their friable texture (i.e., easier to work with wood/shell/lithic tools). Maize root systems can extend 1-2 meters. Being close to the water table reduces the impact of drought on annual yields. For bottomland farming this ironically minimizes the impact of the more commonly assumed yield determinant – rainfall.

Legumes like the common bean (*Phaseolus vulgaris*) can maintain a symbiotic relationship between their root systems and nitrogen fixing bacteria in the soil. The bacteria concentrate in root nodules and provide nitrogen to their hosts. Legumes (e.g., beans, peas, clovers, vetches, and alfalfas) have a minimal need to absorb the nutrient from the soil. But unless the plant is incorporated into the soil (i.e, plowed under) while green (as a so-called green manure) the nitrogen will be lost to the harvest and subsequent burning. The actual benefit of legumes lies in their ability to be planted among other nitrogen consumers. The aboriginal practice of sowing beans with maize was adaptive because it minimized field size and provided a support (the maize stalk) for the climbing legume. The practice, however, does not provide any nutrient value to the maize plants and it would not mitigate nutrient depletion (Gardner, et al. 1985:133; Russell and Russell 1973:359). Conversely, it has been shown (Bowman and Crossley 1911:97) that growing cow peas between the rows can reduce corn yields by as much as eight to ten bu/acre. Munson-Scullin and Scullin (2005) noted in their three year experiment that maize plots with beans had lower yields than monocropped plots. Finally, the most compelling counter argument is *Phaseolus vulgaris* has been shown to be a very poor nitrogen fixing crop (Piha and Munns 1987). Like maize, *Phaseolus vulgaris* benefits from added nitrogen.

Allowing the soil to rest for a period of years would slowly return a field to a higher nitrogen balance largely as a result of naturally raising soil organic matter (SOM) levels. These abandoned areas were referred to as 'old fields' by Native Americans and were distinguished by plants adapted to nutrient depleted conditions (especially wild strawberries [Adair 1930:439]). Short-term fallow periods are common in the tropics' heavily documented slash and burn systems. But in the tropics re-growth and decay occur at a faster rate than in the temperate climate of the southeastern US How long a fallow period is needed to replenish a Mississippian society's exhausted soils? Modelling the process of soil recovery under temperate conditions following continuous agricultural use is a difficult problem. Heidenreich (1971:190) estimated that over 60 years would be required on sandy soils. Sandy loams were expected to replenish themselves after 35 years. Green (1980:224; Likens, et al. 1978) expects a 60 to 85 year period would be required to return an agriculturally disturbed area to a mature secondary stand. However, the Rothamsted experiments suggest that 'old arable soils' would require 100-150 years to raise the nitrogen level from a nearly depleted 0.11 percent to a grassland level of 0.25 percent (Russell and Russell 1973:324). Measurements of natural (i.e., after a long fallow period) soil recovery on prehistoric Mimbres fields in New Mexico suggest that after eight centuries full restoration has not been achieved (Sandor 1995; Sandor and Eash 1991; Sandor and Gersper 1988). Based on these observations, a 125 year recovery period or unassisted 'fallow' would seem reasonable for temperate forest environments.

Maize Agriculture

The success of an agricultural system is defined in terms of sustained yield. There are two parameters that combine to set our yield expectations for any given grain: 'yield potential' and 'potential yield.' Evans and Fischer (1999) make the following distinctions between the two. Yield potential is the maximum achievable yield for a specific cultivar grown free of yield-inhibiting stress and cultivated using optimal techniques under optimal conditions. By eliminating concerns for environmental stress it becomes an idealized measure of the genetically defined productivity of the cultivar (i.e., varietal yield). It is an extremely hard measure to calculate given the difficulty in eliminating stress factors (effects of moisture, temperature, pathogens, insects, lodging, weeds, etc.) and a variety's possible genetic propensity to resist stress. It is the perfect measure to compare cultivar varieties independent of growing conditions – it is the best yield one should expect. For archaeologists it is best applied to comparisons between varieties over the selective breeding history of maize, from insignificant yielding Teosinte to today's Southern Dents.

Potential yield is the measure of a cultivar's expected maximum yield under favourable environmental conditions. It is region and farm specific and generally is derived from computational modelling and simulation. It is useful in comparing reasonable expectations to actual annual yields on a farm by farm basis. But a more realistic measurement, referred to as 'attainable yields' (Duvick and Cassman 1999) can be calculated under intensive agronomic field conditions. Although these input levels are not cost effective on a large scale, this measure is useful as an estimate of possible yields for a new variety or as promotional support for new equipment, techniques, or petro-chemical use. I prefer to use the concept for this study and will couch it in terms of culture specific 'intensive agronomic practices.'

Research involving modern yield measures suggests one important conclusion: the Green Revolution (after A.D. 1960) improved yields largely by enhancing stress resistance among varieties (Duvick and Cassman 1999; Evans and Fischer 1999). Controlling for other factors, there has been little improvement in yield potential as defined above. For maize the most important improvement was increased tolerance of higher plant densities, i.e., more plants per unit area. This contrasts with prehistoric genetic manipulation where, over thousands of years, the goal was improving varietal yield potential. Throughout North America at any given time cultivar choices were limited. As an example, the ancestors of modern Corn Belt dents were formed by crossing Northern Flints with Southern Dents (Brown and Anderson 1947, 1948). The earliest date for such mixing is probably around A.D. 1840, because it was then that we find evidence of 40 maize varieties of various racial origins (Bowman 1915:3). This represents an eightfold increase in variability over the five varieties (four flints and one dent) known to have existed in 1814 (Bowman 1915:3). For our purposes this is significant because it implies yield potential differences between observable ethnohistoric yields and those of the late prehistoric should be small. This provides us with an agronomic basis for setting yield potentials for Mississippians. But can attainable yield measures be defined? Would they be constants, like yield potential, or variable over time? To answer these questions we must examine the factors determining attainable yields.

US Corn Belt farmers during the 19th century planted maize much the way their ancestors and Native Americans did centuries before. Open pollinated varieties would have been comparable to late prehistoric choices (hybrid varieties would not become widely accepted until the 1940s). Three to four seeds were planted in 'hills' spaced approximately one meter apart in each direction. Soil would be mounded around the young stalks forming a hill, which helps to stabilize the top-heavy plant in high winds. The resulting plant densities would be 3556 hills per acre (8788 hills per hectare) producing between 10,668-14,224 plants per acre (ppa) (26,360-35,147 plants per hectare [pph]). Early 20th century experiments concluded that densities of four plants per hill was optimal, as greater numbers of plants induce too much competition for water and nutrients (Bowman 1915:166). In comparison to modern practices, the average pph for four Corn Belt states (Indiana, Iowa, Illinois, and Minnesota) was 47,400 (19,180 ppa) in 1973 (Larson and Hanway 1977:645). In general, late 19th-early 20th century research is very applicable to our needs.

Crop agriculture (as well as horticulture) maximizes the process by which plants extract nutrients from the air, soil, and water, converting them into food products which can be harvested, hauled away, and consumed. This 'hauling away' begs the question: How are the nutrients returned to the environment for succeeding plantings? The simple answer is they are not, at least not at a rate equal to or exceeding their removal. High protein grains are built from translocating nitrogen from the soil to the plant and then to the seeds. The net result is a loss of nutrients for next year's crop. Modern (i.e., post A.D. 1940) techniques incorporate petro-chemical and mechanized subsoiling solutions to offset the imbalance. Externally derived fertilizers must be added to maintain production over a continual long term. Alternatively proper crop rotations, though less optimal in the short term, do minimize the rate of loss over short periods.

The negative impact of long-term, continuous field cropping systems has been noted for Colonial America (Carman, et al. 1964; Cronon 1983; Donahue 2004:203-208) as well as 13th-15th century Europe (Campbell 2000; Campbell and Overton 1991). To empirically estimate the soil's response to Mississippian continuous cropping, one needs to duplicate their practices over many lifetimes. Fortunately agronomic researchers have provided a suitable substitute. Agricultural research stations dating to the mid-19th century in England and the United States have demonstrated the negative affects of repeated plantings of the same crops, without fertilizers, year after year (Brown 1994; Buyanovsky, et al. 1997; Darmody and Peck 1997; Friedman 1996; Hendrix 1997; Jenkinson, et al. 1994; Powlson 1994; Rasmussen, et al. 1998). These long-term crop studies, inspired by the Rothamsted plots (Hall 1905, 1917), provide controlled experiments on yield factors for a wide range of crops. Today's agronomists take fertilizer application for granted – it has been a given on most US farms since the late 1940s – and their research focuses on demonstrating higher yielding, more sustainable practices. But a century ago this was not the case. Test plots were designed to demonstrate and encourage best practices for their times. Our interest lies with those comparative plots used to demonstrate the value of fertilizer or crop rotations on maize production. The experiments consisted of continuously planting maize on the same plots for a number of years (some over a century). Some did not receive added amendments while others were treated with manure, chemical fertilizer, and/or crop rotation practices. A clear recognition of nutrient exhaustion is possible by comparing their annual yields over time.

Several research institutions started long-term maize demonstration plots. Among these were the Morrow Plots at the University of Illinois (first in the US over 120 years ago), Sanborn Field of the University of Missouri, Ohio's agricultural field station plots at Wooster, and Auburn University's Cullar's Rotation. Yield data for the Morrow Plots (E.D. Nafziger, personal communication, 2005), Sanborn plots (Brown 1994; Buyanovsky, et al. 1997), Wooster plots (Weir 1926), and Cullar's Rotation (Thomas Foster, personal communication; Foster n.d.) are available for this study. To contrast the long-term soil response to continuous cropping, Figures 6.1 and 6.2 present the cumulative yields (bu/acre) of each set of plots (treated vs. untreated) over the duration of the experiments. To maintain approximate consistency with late aboriginal varieties only yields prior to 1941 were examined (the likely time of hybrid introductions). Because continuous cropping degrades the soil's ability to maintain yields over time, plotting the cumulative yields provides clearer evidence of the divergence in attainable yields. Interestingly, linear regression results (Table 6.1) on the untreated plots produced similar slopes 16.90 bu/yr (1061 kg/yr) for Sanborn's and 16.75 bu/yr (1051.6 kg/yr) for Wooster's plots compared to their treated plots' 31.34 (1967.6 kg/yr) and 37.90 bu/yr (2379.4 kg/yr), respectively. The trends for the treated plots support earlier arguments that the expected yield potential for late aboriginal maize should be close to 30 bu/acre.

These yields can be recalibrated to reflect a late prehistoric yield potential of 30 bu/acre (1883.4 kg/ha) by assuming the maximum observed yield (MAXy) for the treated plots was equivalent to a prehistoric yield of 30 bu/acre. This is necessarily a conservative estimate of the various plot varieties' yield potentials. The ratio of 30/MAXy multiplied by each observed yield produces an estimate of the probable observed yield for late prehistoric maize varieties at any year, t.

The ratio of cumulative untreated to treated yields (f_t), is inherently nonlinear over time. As a function of time (t), the ratios can take the form:

$$f_t = a(t^b) \qquad t \in [1, 2, 3, \ldots] \quad (6.5)$$

Nonlinear regression analysis on the four sets of observed ratios produces the equation parameters shown in Table 6.2. These in turn produce, for any year t, an annual maximum attainable yield function of:

$$Y_t = 30f_t \text{ (bu/acre)} \qquad (6.6)$$

Figures 6.3 and 6.4 display the attainable yield limits and the adjusted observed yields of each untreated plot. The curves represent an upper bound for annual yields. Very few observed yields equalled or exceeded their limits. The Cullar's Rotation is unique in its irregularity and extremely low untreated yields (Figure 6.4 C). These plots are on highly permeable soil where leaching of nutrients and organic matter would be rapid. By the late 19th century they were probably already at the extreme limits of continuous production and far beyond the levels developed at the other research stations. Notably, the Cullar's soils are most like those expected to be associated with most southeastern Mississippian societies.

It is this exponentially decreasing attainable yield function that creates the agrarian dilemma: How can harvests be maintained when the act of farming reduces attainable yields each year? Yields are not simply the result of annual weather patterns. Although rainfall and temperature extremes contribute to actual yields, they do so *under* the attainable yield curve. Soil nutrients dissipate under agrarian practices.

Nye and Greenland outlined six causes for the documented decline in production of shifting agricultural systems:

1. Deterioration in the nutrient status of the soil
2. Deterioration in the physical condition of the soil
3. Erosion of the top soil
4. Changes in the numbers and composition of the soil fauna and flora
5. Increase of weeds
6. Multiplication of pests and diseases. (1960:75)

The deterioration of nutrients and the physical condition of the soil, top soil erosion, and loss of soil fauna and flora combine to reduce the viability of the growing medium. Nitrogen (N), phosphorous (P), and potassium (K) are the three critical elements absorbed from the soil by maize (Larson and Hanway 1977:634). Deficiencies in these elements generally lead to a decreased growth rate and stunting. Nitrogen deficiency results in barren ears, stunted kernels, and poorer quality proteins. Phosphorus deficiency can minimize successful pollination by delaying silking. Potassium-deficient plants tend to produce small and poorly filled ears as well as weak stalks susceptible to rot (Larson and Hanway 1977:635-636).

Absorption of N is dependent on the processes of oxidation (NO_3) and reduction (NH_4) largely as a result of nitrifying bacteria (Gardner, et al. 1985:110). Because these are biological processes, they are easily affected by temperature, moisture, and soil pH. Nitrification is minimal during the cold, wet months of winter and spring and optimal in well aerated soils when the temperature exceeds 25°C. Denitrification becomes a problem under warm, waterlogged conditions and during leaching when the soil is well aerated. Late-successional vegetation tends to produce nitrification inhibitors (tannins and phenols) which are slowly removed by leaching during cultivation.

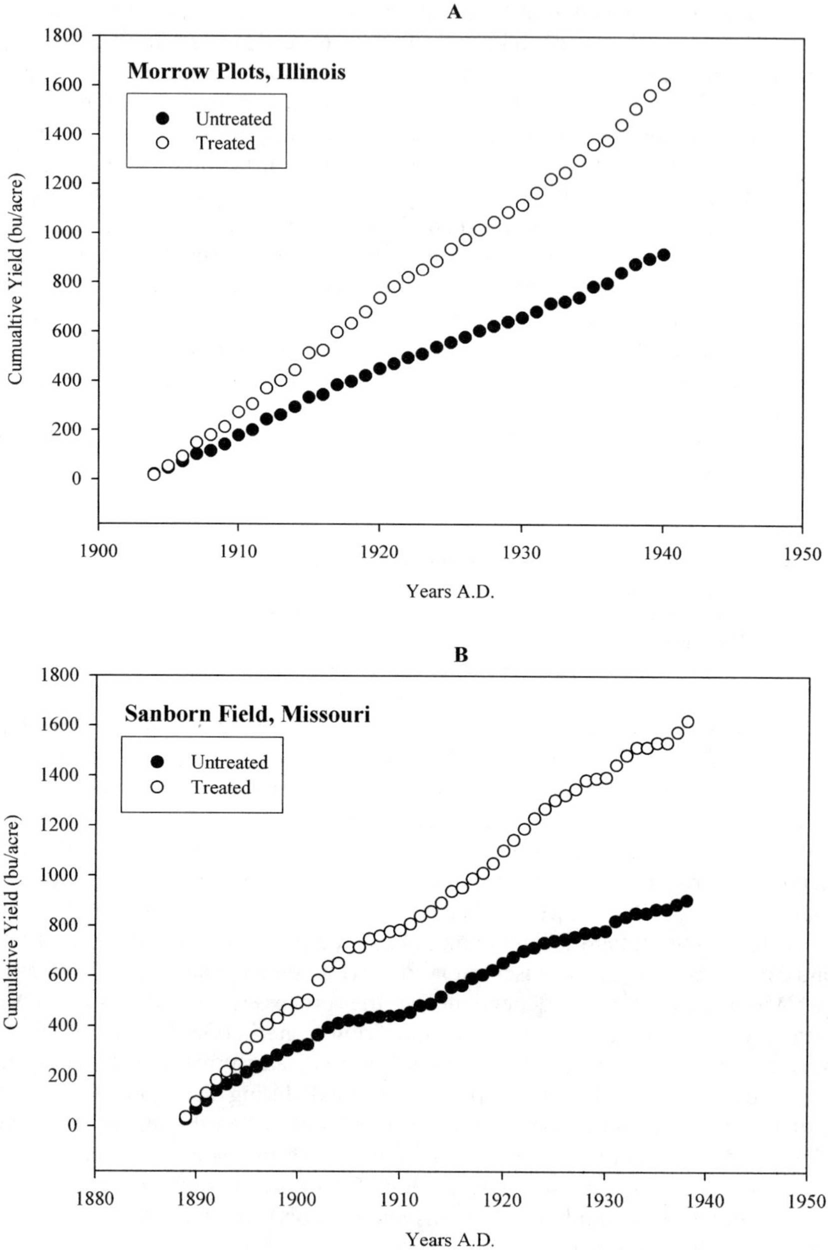

Figure 6.1 Cumulative maize yields on treated and untreated plots at Morrow Plots and Sanborn Field

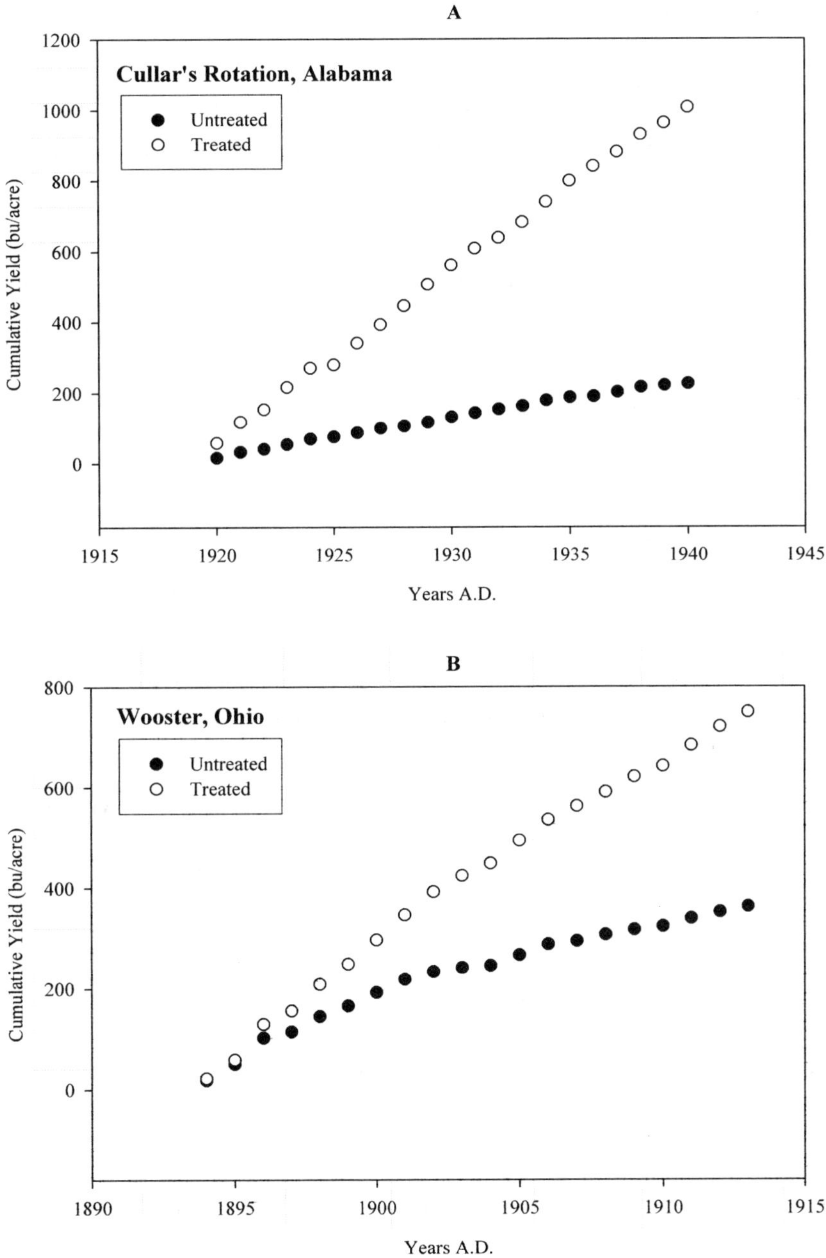

Figure 6.2 Cumulative maize yields on treated and untreated plots at Cullar's Rotation and Wooster's Experimental Station

Table 6.1 Partial linear regression results on long-term, cumulative yields, Yield = *b*t + c

Plots	Soil	Years A.D.	Status	*b* (bu/acre/yr)	R^2
Cullar's Rotation	Loamy sand	1920-2001	Treated	47.957	0.998
			Untreated	10.589	0.997
Morrow Plots	Loess over prairie	1888-1940	Treated	43.501	0.999
			Untreated	24.376	0.997
Sanborn Field	Silt loam	1889-1938	Treated	31.337	0.993
			Untreated	16.904	0.988
Wooster Plots	Silt loam	1894-1913	Treated	37.898	0.991
			Untreated	16.745	0.953

Table 6.2 Nonlinear regression results for the ratio of cumulative yields, $f_t = a(t^b)$

Plots	a	b	R^2	MAXy (bu/acre)
Cullar's Rotation	0.294	-.083	0.863	63.1
Morrow Plots	1.027	-.180	1.0	73.6
Sanborn Field	0.822	-.101	1.0	79.4
Wooster Plots	0.926	-.204	0.920	70.57

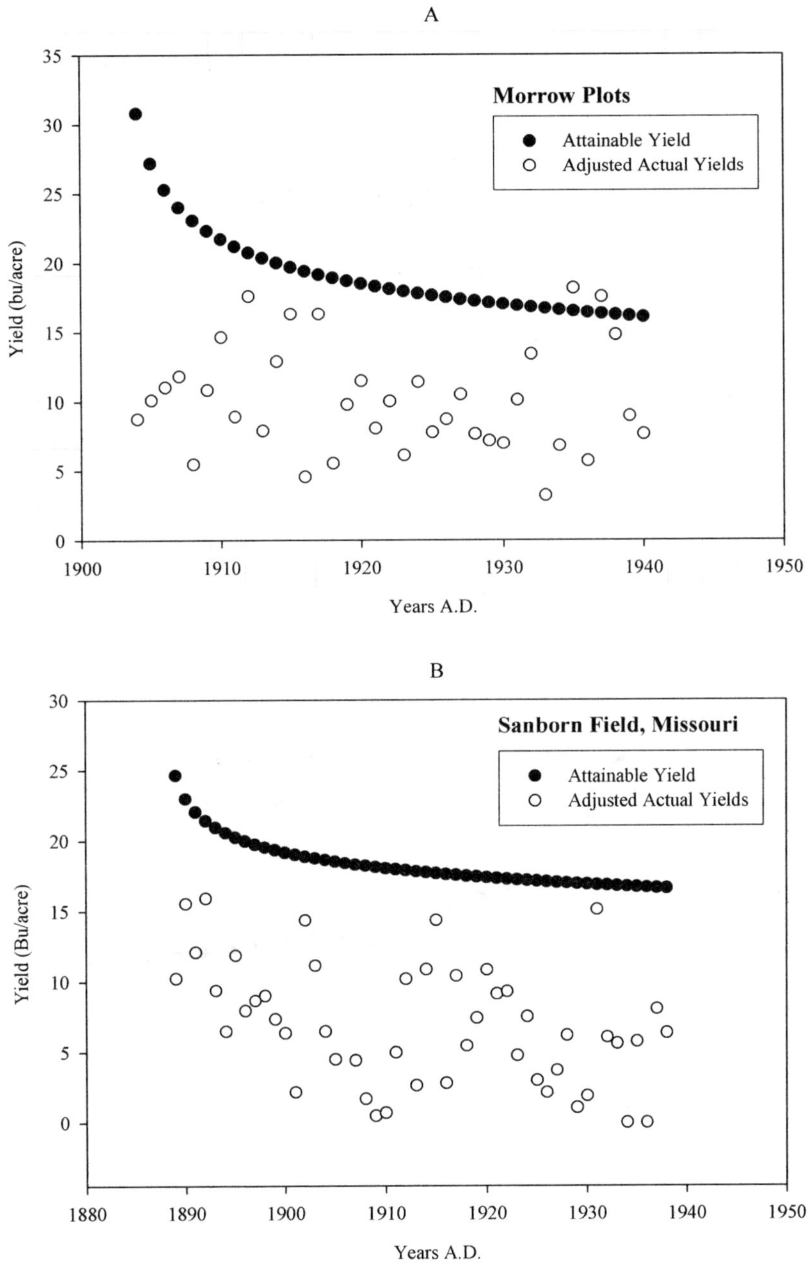

Figure 6.3 Estimated attainable yields for Morrow Plots and Sanborn Field

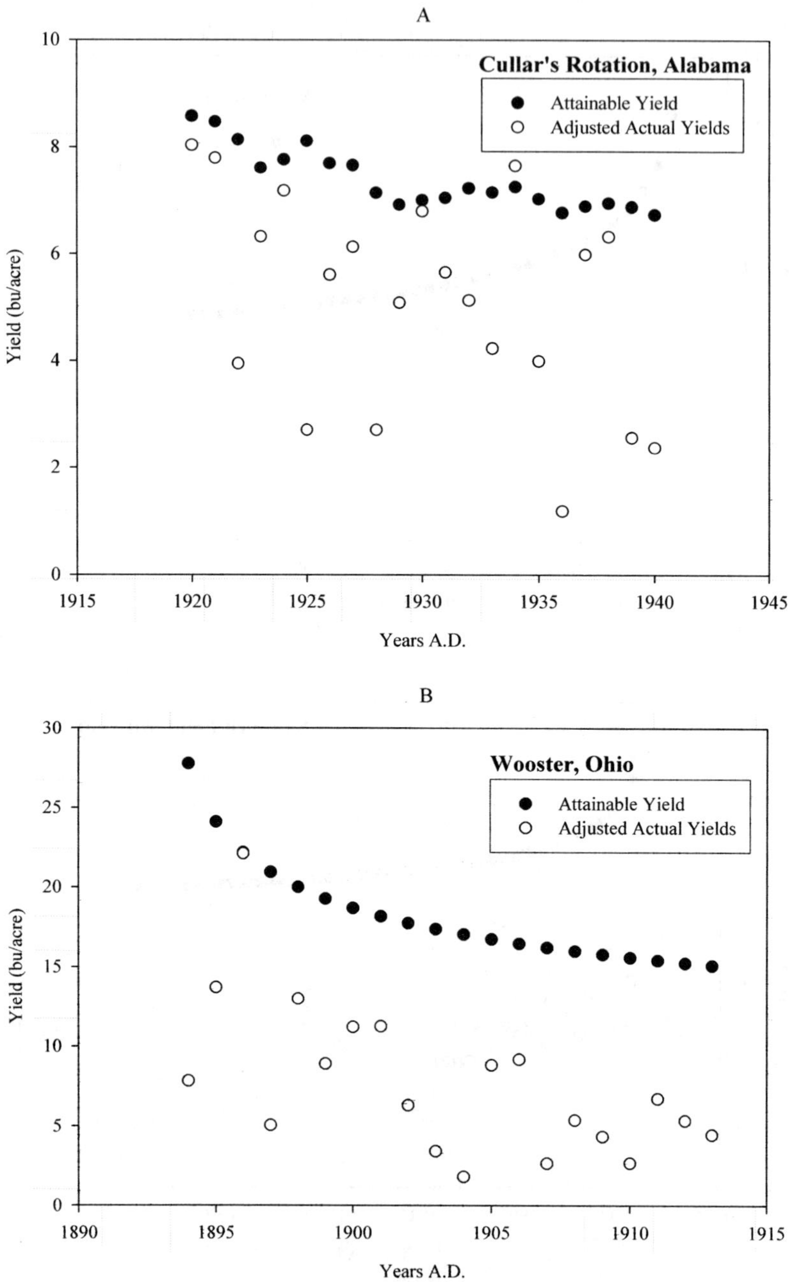

Figure 6.4 Estimated attainable yields for Cullar's Rotation and Wooster's Experimental Station

Phosphorus is represented in both organic and inorganic portions of the soil matrix. Most P absorption is dependent on the element being in solution, which accounts for the smallest share of soil P. Although the concentration of soluble P can be extremely low, root action results in plant levels up to 1000 times that of the surrounding soil. Thus, plants can quickly incorporate most of the available P (Gardner, et al. 1985:115-116). Potassium is primarily derived from minerals, especially clay minerals like montmorillonite. Although only about 1 to 3 percent of the total K in soil is available through exchange or solution, most soils are sufficiently buffered to sustain constant levels from year to year. Like N, potassium absorption is optimal at 25°C (Gardner, et al. 1985:117-118).

Rates of nutrient uptake vary according to the growth stage of the plant. Potassium absorption usually is complete by the time of silking, while N and P continue to be incorporated until the plant is almost mature. Through the process of translocation, N and P are largely (66.7 to 75 percent) concentrated in the grain by harvest time. Potassium, on the other hand, tends (75 percent) to remain in the leaves and stalk (Larson and Hanway 1977:634-635). Using a standardized 10,000 kg/ha (159.3 bu/acre) we should expect 200 kg of N, 36 kg of P, and 190 kg of K to be incorporated in the grain and stover (bulk plant remains) of modern Corn Belt varieties (Larson and Hanway 1977:634-635).

What are the essential contributors to nutrient degradation? Of the three primary elements, nitrogen deficiency is the most critical. Only about 1-3 percent of the predominant organic N is mineralized (i.e., converted to nitrate and ammonium) each year. In Native North America nutrients were further depleted each year when fields were burned prior to planting. The burning killed microbes and released 95 percent of available N and 54 percent of the potassium stored in plant remains (Arianoutsou and Margaris 1981). Nitrogen loss is also strongly associated with the decline in SOM under continuous farming and burning. SOM provides the fuel for nitrogen fixing organisms through the decay process. Without continued subsoil incorporation of organic material, nitrogen production cannot be maintained indefinitely. So, nitrogen is volatilized by burning, leached out of the soil by rainfall, and harvested in the protein constituent of the grain. Nitrogen is clearly the first nutrient to reach deficiency levels.

Weed and insect infestations, along with disease epidemics, will tend to exacerbate reduced yields at the end of the viable life span of a field largely as a result of nutrient deficiencies. Further, reduced yields fail to meet the nutritional and caloric needs of populations in more than just quantitative terms. Reduced nitrogen produces poorer quality proteins (Uribelarrea, et al. 2004). In the absence of systematic rotation, fields would produce until some point when their yield per unit labour is insufficient to support society's needs and field abandonment would occur. Most significantly, fields that have become depleted have been shown to require 100 to 150 years to naturally replenish their organic nitrification potentials (Russell and Russell 1973:324). For modelling purposes, the use-life of a field can therefore be directly related to the attainable yield curve.

As a result of these conditions, Mississippian agriculture should have produced fluctuating energy/nutritional situations which periodically triggered self-organizing responses from its parent populations. To better understand the boundary conditions of Mississippian agriculture we need a model of maize production dynamics from society's perspective. In particular we must identify critical points affecting the agrarian system's ability to export a constant, predictable flow of energy into the parent culture. At these points, in response to cultural restraints and inputs, the maize system creates its own new order, often resulting in undesirable lowered attainable yields (energy production).

Is it possible to associate agrarian instabilities with observed cultural transitions? This would require calculating continuous agricultural potentials, P, much as Meggers argued. When the varying rates of change, dP/dt, are close to zero, stable, nearly steady state, energy flows should be expected. Whenever $dP/dt \neq 0$ instabilities would have occurred that could have triggered cultural changes in the parent culture. One measure of such potential could be the amount of available, undepleted, arable land. For this study P will represent this amount of land, over time. Unfortunately, unlike Adams' Victorian examples, we lack written records for our study populations. A simulation is needed to explore the interaction of these agronomic dynamics and Mississippian behavioural variables.

The Simulation

A generalized simulation of a land reserve of 20,000 ha for a fixed population of 3000 was executed 100 times for the years A.D. 900-1700. The expected yield potential for the maize varieties was 30 bu/acre (1883 kg/ha) and attainable yields were expected to annually be reduced following Eq. 6.6 as arbitrarily defined by the Sanborn Field data. To offset low yields, field sizes were allowed to increase up to 0.4 ha/person. Three consecutive crop failures (i.e., yield plus the previous year's surplus fails to meet demand) constituted a depleted field. Once depleted the field was abandoned for 125 years after which it was returned to the overall land reserve. To account for minimal hoeing and the resultant heavy weed growth, all yield equations used 2t for t (i.e., twice the consumption of nutrients). Demand was annually estimated based on Eq. 6.4. The simulation also examined the impact of behavioural choices. In addition to setting the size of the fields per person, the population could fission (reduce its numbers to an acceptable levels) if the available land was not sufficient to support the existing population, or the population could abandon the valley if all lands were depleted. The fixed population size of 3000 was selected because it was the largest population that did not fission during the 801 years.

Yields would be expected to fluctuate based on climatic conditions. Simulated yields were calculated using a random normal distribution with adjusted (30 bu/acre yield potential) mean of 12.47 bu/acre (782.87 kg/ha) and a standard deviation of 6.62 bu/acre (415.60 kg/ha). Climatic specifics are becoming more available as tree ring studies contribute to our understanding of weather patterns during Mississippian periods in many southeastern regions (e.g., Stahle and Cleveland 1994). However, strong correlations between climate (especially rainfall) and expected maize yields have not been demonstrated (Anderson, et al. 1995) and may not be warranted (Albrecht 2000). Estimates of rainfall amounts alone are not sufficient to accurately predict maize yields. Although continued consideration of the potential of including this information into this model is warranted, the complexities of this undertaking were not dealt with here. For our purposes it is sufficient to simulate yields acknowledging fluctuations as a partial response to temperature and rainfall. But the impact of minor droughts would often be mitigated by the terrace's proximity to the water table.

Figure 6.5 represents the amount of land available each year as an annual average of the 100 simulated runs. The initial 100 years are marked by dramatic shifts in field size as the *per capita* demand for maize increased. To meet growing demand, Mississippian technology was limited to four options: increasing field sizes, increasing plant densities within the limits of the varieties, moving to new fields, and/or improving the varietal component (i.e., breeding better varieties). We can be fairly certain the latter option was important, initially, in the development of flint varieties from *Maiz de Ocho*. However, there is no evidence of significant improvements in the productivity of these varieties until early in the 19th century. The simulation allowed field sizes per person to increase to a limit of 0.4 ha/person. After three consecutive crop failures a field was moved and, if needed, the size was increased. Mathematically the first hundred years was an unstable time in terms of defining agricultural parameters. The instability would have had an impact on social relations as labour was organized and allocated to fulfil demand. The duration of this period corresponds well with the Emergent Phase of Mississippian development.

Between A.D. 1000 and 1300 the available land reached a balance between increasing demand and availability of 'old fields' sufficiently stabilized to be used again. This was a stable period correlated with the general time period of the Middle phase of Mississippian development. The years following were unstable and associated with the Late Mississippian phase. Again, agrarian instabilities correlate well with evidence of cultural upheavals and deteriorating health at the close of the Mississippian Period. After A.D. 1450 stability returned but at a dangerously low potential. This represents the pre-contact to contact period of the late prehistoric in the Southeast. Briefly, the points of change (discontinuities) mark transition points (phase shifts) that correlate well with general, observed Mississippian phase transitions.

Discussion

Anecdotal evidence documenting the practice of grain agriculture over the past few centuries combined with the early systematic study of agronomy clearly supports the axiom – crop agriculture is not a self-sustaining endeavour. Taken alone this may seem obvious or even frivolous, but the archaeological literature seldom acknowledges it. Though rare, ethnohistoric studies by anthropologists have documented the impact of soil exhaustion on the settlement patterns of historical groups, such as the Iroquoian populations (McGlade and Allen 1986; Pendergast 1996; Tooker 1964). Charles Redman (1999:98-99, 122-126) has discussed agriculture's role in accelerating soil depletion (nutrient and erosional) and the resulting impact on cultural development, especially in terms of settlement systems and labour intensification. Yet it is more common for archaeologists to refer to centuries of continuous prehistoric agrarian occupations with no appreciation for the historically obvious implications of such a scenario. The notion that attainable yield potential is a constant, self-maintained quantity is clearly an archaeological myth like some form of perpetual energy. If long-term field sustainability is questionable, the concept of 'permanent settlements' must be reworked to incorporate a more shifting, transient organizational system. Understanding the dynamics of attainable yields is critical to understanding the impact of agrarian systems on emerging social adaptations and settlement system adjustments. Further, the implications of non-sustainability are such that no one should consider climate change as the only limiting factor in agrarian economies.

Reworking these observations under a dissipative structure paradigm is one way of incorporating these processes into an understanding of emergence. From the perspective of the inherent dissipative structures embedded within maize agricultural systems, we can identify one major source of fluctuations: societies demanding higher yields than the eco-agrarian system can support. In response to external human actions the attainable yield potential is lowered to levels consistent with the soil's available nutrient capacity. Without external inputs to offset the losses (e.g., fertilizers) yields would decrease to significantly lower levels, though never permanently going to zero. Prehistoric societies were not capable of significantly replacing what they extracted. One recourse involved moving on, letting natural processes restore the land. Another would be intensifying labour and land utilization in an effort to obtain sufficient returns on low yields.

From the perspective of the parent society, the drops in attainable yields are seen as repeated fluctuations that threaten to reduce the contribution of d_eS to Eq. 6.3. To maintain Eq. 6.3 at steady state levels, socio-political adjustments would be needed to intensify agricultural efforts and establish elaborate land management practices. One practical example is the adjustment in field size. An increase in field size per person translates into an increase in labour costs. Alternatively, abandoning fields and village sites requires an extensive land management system that may span valleys. Resolution of these issues clearly falls under the self-organizing aspects of cultural dissipative structures. Far from equilibrium, continued system maintenance requires the creation of more complex mechanisms to restructure or divert stress. But once choices become limited, the society, like the exhausted fields, may need to devolve to a lower total dS state which generates a concomitantly lower d_iS in scale with a depleted d_eS.

This model of agricultural potential is largely inspired by the conceptualization of dissipative structures. Simulations similar to this can be used to define the timing of fluctuations in the extraction of d_eS (not a constant). The fluctuations can, in turn, be compared to observed archaeological evidence in an

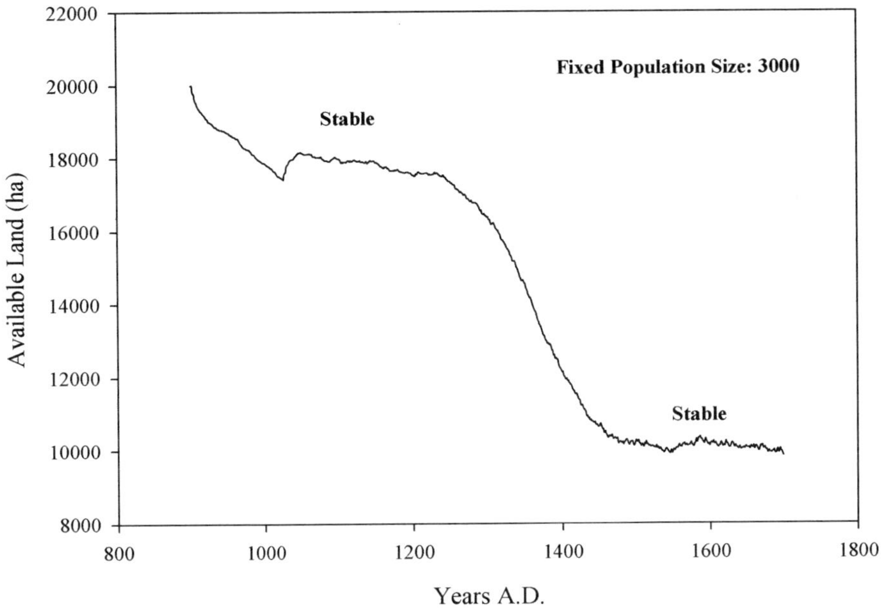

Figure 6.5 Average available land from an initial 20,000 ha base (n=100 simulations)

attempt to correlate their timing with the creation of new order as predicted by dissipative structures. The structures also contribute to the definition of fitness. The thermodynamic premise then enhances detailed agent based simulations that can explore the intricacies of the internal, cultural dissipative structures.

The dissipative structures paradigm is not without critics (Bricmont 1997; Jaynes and Rosenkrantz 1983). Many concerns relate to the concept of irreversibility (processes associated with d_iS not discussed here; see Nicolis and Prigogine 1989:61-65) and microscopic application of the theory. I follow Adams' advice and acknowledge these concerns while pressing forward in the use of dissipative structures as one of many mechanisms of emergence, taking advantage of what their perspective offers. There is, after all, still some intuitive value in considering culture in nonlinear, thermodynamic terms.

Chapter 7

Approaches to Modelling Archaeological Site Territories in the Near East

T.J. Wilkinson
University of Edinburgh

Introduction

The origin and development of systems of human settlements has been of interest to students of complex adaptive systems for virtually as long as such systems have been studied. Particularly prominent has been research on settlement in the American Southwest where Timothy Kohler and colleagues have been successful in producing long term simulations of settlement in the context of an exceptionally fine-grained run of climate proxy data provided by dendrochronology (Kohler, et al. 2000). However, such studies have entailed the modelling of relatively small scale settlements in the New World: no attempt has yet been made to model urban scale settlements in their area of origin, namely the Near East. This chapter will therefore shift the focus of attention to the Near East, specifically the zone of rain-fed cultivation in Upper Mesopotamia, where large urban-scale communities developed from as early as the fourth millennium B.C. The primary focus will be on one aspect of the modelling of ancient settlements, namely how Near Eastern settlements have been analyzed within their territorial context in the past, and how it is possible to develop more advanced models that can be contextualized within a political economy and social framework in a non-deterministic manner.

Because any human settlement must be sustained by food production and other forms of economic activity, it is necessary to analyze archaeological sites within an economic territory within which the main locus of economic activity was concentrated. In the past this was undertaken from the perspective of site catchments (as discussed below), but if we are to undertake a realistic accounting of the political economy, catchments must be extended to encompass a much larger area than that which supplied subsistence provision. First it is necessary to trace how settlement catchments were initially modelled using traditional methods, and second I suggest how agent based models can supply a rich framework of analysis that can result in more realistic, and often nonlinear results.

In recent years there has been a discernable decline in the analysis of settlements and their economic territories, a lull that has been ascribed to the rising influence of Post-Processualism so that 'the environment has gone back to being a rather passive backdrop to a world created in the human mind' (Bintliff 2000:21). Nevertheless, because the linked settlement-territorial system continues to provide a crucial context for understanding human social development, there is now renewed activity in this area, especially thanks to the increased use of agent based models. To provide a historical context for the main part of this chapter, this brief review starts with site catchment analysis, continues with the estimation of site sustaining areas, and then reaches the logical development of these techniques, in which landscape archaeology is employed to provide estimates of site populations and agricultural areas. With every developmental stage of site territorial analysis we see both increasing sophistication of technique, but equally the incorporation of a broader range of assumptions. Although providing greater complexity of analysis, these developments equally result in a wider range of variability as an outcome of the modelling. This review therefore takes the next logical step in territorial analysis by discussing how agent based modelling can be employed to tackle site territorial analysis, and specifically how this injects dynamic behaviour and perhaps evolution into the resultant models. This new generation of models should provide a much richer level of analysis than was ever possible using conventional 'static' techniques.

Traditional ways of modelling site territories have long been criticized for being unrealistic because: a) they are environmentally deterministic; b) they make unwarranted assumptions about population densities; food consumption, etc.; c) they do not allow for what may have been fundamental features of the ancient economy such as feasting; d) they assume a homeostatic, equilibrium situation; and e) they do not allow for human agency. Moreover, and of relevance to the present volume, each methodology enshrines within it a set of linear or deterministic assumptions that do not allow for the incorporation of the idiosyncratic or surprise features that are so common in the real world. Taken together, the foregoing represent a fairly significant set of criticisms, but equally, traditional techniques have been productive by allowing archaeologists to understand the development of ancient economies and moreover, they have taken the comparative analysis of ancient settlements out of the arena of the site, to the level of the region. However, with the development of a new range of computer-driven techniques, it is now possible to proceed with quantitative approaches to the analysis of ancient villages, towns and cities within the framework of the regional economy.

The following brief history is only intended to provide a basis for the discussion of agent based modelling that follows. For a recent more detailed analysis of the field of site and territory see Bintliff (1999b, 2000).

Site Catchment Analysis

The methods of site catchment analysis were a staple feature of processual archaeology during the 1970s and the 1980s (Flannery 1976; Jarman, et al. 1972). Drawing on a long tradition of geographical research, extending back to the German economist Von Thunen, and building on Chisholm's classic book *Rural*

Settlement and Land Use (Chisholm 1962), site catchment analysis explicitly places archaeological sites within their soil landscape, conventionally defined as that land within one hour's walk of the site (for sedentary communities ca. 5km), suitably weighted according to distance (Figure 7.1). Using the traditional pattern of land use as a proxy for that of ancient times, this methodology was both inferential and environmentally deterministic. In other words, the nature of the ancient subsistence economy was inferred through the patterns of traditional agriculture, and little new archaeological data was introduced to make an independent contribution to the interpretation. Hence, there was a tendency, for example, to assume that if a site is surrounded by a cultivated plain today, the component soils would have been cultivated in the past. Therefore the site economy would have been inferred as being predominantly agrarian, with a significant emphasis on the production of staple crops. Although such assumptions may operate for large areas of the Near East and Mediterranean, this is not a universal norm, and in many upland plains, for example in SW Arabia and perhaps the Caucasus, terrain that is cultivated today may in the past have been pastoral lands (Wilkinson 2003:184-209). If that were the case, then the drawing of cultivation catchments around sites which are cultivated today may be seriously misleading.

Although endorsing site catchment analysis Kent Flannery argued that it was necessary to make certain amendments in order to make it a permanent feature of the archaeologist's arsenal (1976:95). Such modifications extended the relevance of site catchment analysis; nevertheless, criticisms along the lines noted above have resulted in the techniques of site catchment analysis being applied with decreasing frequency in recent years.

From Settlement Patterns to Site Sustaining Areas

Despite the relevance of site catchment analysis to archaeological survey, many of its first applications were not to regional surveys but to individual sites. With the increased application of archaeological survey since the 1970s there was increased awareness of the need to analyze areas beyond the site, even though it was difficult to do this using the techniques then available. In the Near East, the estimation of agricultural areas by means of site sustaining areas was pioneered by David Oates (1968) and Robert McC. Adams (1981). The approach of Adams, which was explicitly applied to data from archaeological surveys, differed from the methods of Jarman, Higgs and Vita-Finzi by incorporating site area data to 'project' the site into its hinterland. During the 1980s Gil Stein and Patti Wattenmaker (1990; Stein 1994) applied similar techniques to employ occupied area (as estimated by the on-site sherd scatter for the period in question) to generate estimates of site population and from these to estimate the cultivated area required to feed that population (Figure 7.2). Given the absence of more detailed data it was necessary to make the assumption that for the period in question, the settled area was occupied at a specified population density (conventionally 100 or 125 persons per ha), a figure derived from ethnographic studies of traditional settlements in the region (Kramer

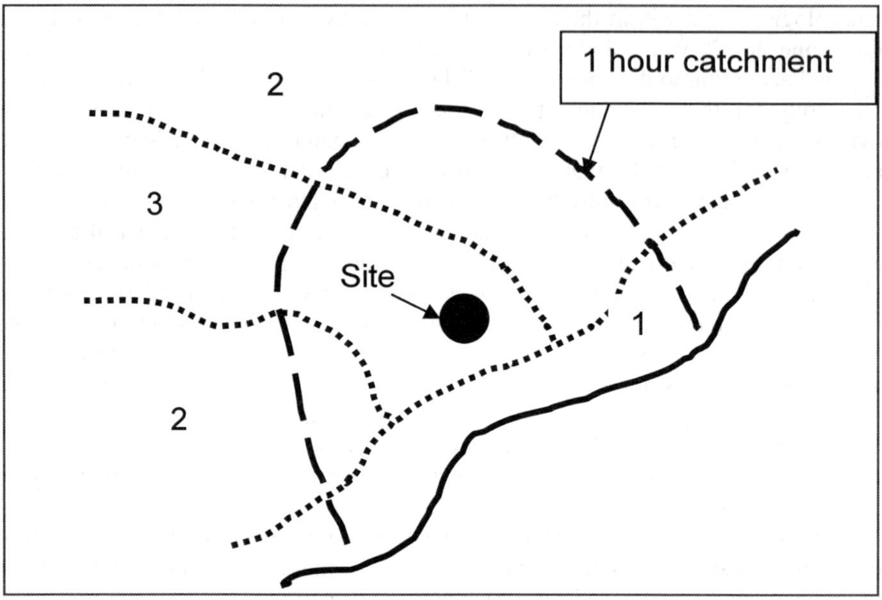

Figure 7.1 Sketch depicting the basic elements of site catchment analysis for a hypothetical coastal site showing different soil resource zones 1-3

1980; Sumner 1989). Overall Stein and Wattenmaker were able to conclude that the political economy of the mid third millennium B.C. was significantly larger in scale than that of the preceding early third millennium B.C. (Ninevite V). Moreover, whereas the earlier Ninevite V satellite settlements fell outside the agricultural catchments of the centres, by the Akkadian period, when the political economy must have attained its maximum scale, satellite communities around the major Akkadian centre of Tell Leilan clearly fell within the sustaining area of that site and were probably within its administrative orbit as well.

This compelling conclusion meant that archaeologists were now able to assess the spatial component of the political economy in ways that were hitherto impossible. Nevertheless, not only did the conclusions depend upon a number of assumptions and simplifications, there was no real cross-check on the sustaining area itself; it was simply generated from the area of the contained site. As a result, there was a tendency to assume that the population of the settlement was necessarily supported by the fields that surrounded the site. Unfortunately it was not possible to make an independent estimate of agricultural areas to cross check these conclusions.

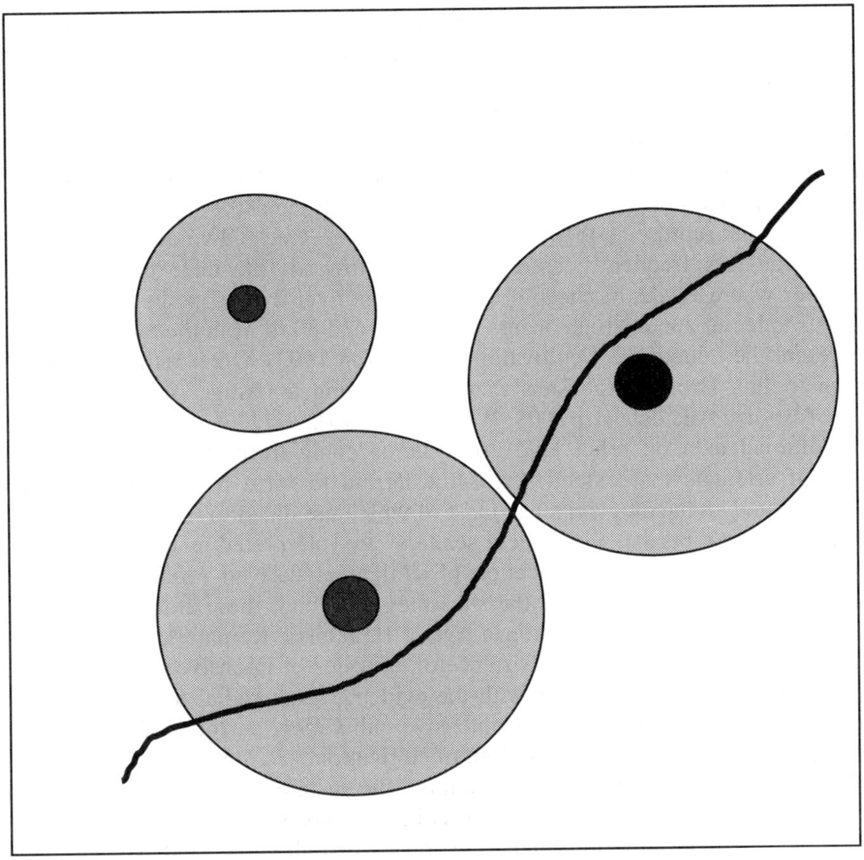

Figure 7.2 Sketch depicting the basic elements of site sustaining areas (shaded circles) in direct proportion to the size of the contained archaeological sites (small black circles)

The Role of Landscape Archaeology

In addition to site catchment analysis and sustaining areas, archaeologists are able to analyze site territories by the use of Thiessen polygons (Ruggles and Church 1996), a method of spatial allocation of territory that is much easier to perform now through the use of Geographical Information Systems (GIS). Nevertheless, this method is arbitrary in its creation of territories, because it fails to use actual archaeological data except for site size and location. Under ideal circumstances Near Eastern archaeological sites can be placed within a regional economic context by the use of landscape archaeology. For example, in parts of northern Syria or

Iraq, one can recognize 'signature landscapes' which enable the archaeologist to construct the pattern of settlements together with their agricultural territories. These ensembles comprise the central, nucleated mounded site (or tell), an outer cultivated zone inferred from landscape data, together with associated trackways and outlying pastures (Wilkinson 2003).

In parts of northern Syria and Iraq, topographic hollows, soil or vegetation marks radiate from significant tells to form a conspicuous feature of the landscape. These appear to represent relict 'hollow way' roads, and the point where these features fade out represents the zone where the village tracks dwindle to nothing as they become less frequently used by humans towards the edge of the village territory or where flocks of sheep or goats fan out over the steppe beyond (Figure 7.3). This fade out zone, which occurs between 3 and 5 km from the tell, is inferred to represent the boundary of cultivation (Wilkinson 1993). Other radial tracks can be seen to link Bronze Age sites thereby providing a vestige of cross country routes of Bronze Age date (Ur 2003; Wilkinson 1993).

Additional data on relict land use systems come from low density off-site scatters of artefacts that extend across the terrain to form haloes of artefactual material around archaeological sites, most of which are dated to the Bronze Age or Hellenistic/Roman period. Such 'field scatters' are interpreted as representing the debris remaining from the application of settlement derived refuse to fields as fertilizer (Wilkinson 1982). Like the sites they surround, these extensive scatters are in most cases datable to the Bronze Age or Hellenistic/Roman periods. Overall, field scatters suggest the former existence of episodes of intensive agriculture and manuring. Such scatters, together with the evidence from radial tracks, provide a valuable picture of an inner zone of intensive cultivation, surrounded by an outer less intensive zone extending to the limit of cultivation.

Under ideal circumstances, by combining the sustaining areas estimated from site areas (as described above) with the cultivated areas derived from landscape archaeology, it is possible to compare estimates of cultivable land with the land required by the population. This demonstrates whether individual sites were self supporting, producing a surplus (presumably to be exported to a nearby centre) or running a deficit, in which case they would be net importers of staple foods. In one particularly fortuitous case, cuneiform tablets from the site of Tell Beydar provided lists of plough animals used to plough the fields around Tell Beydar (third millennium B.C. Nabada). This contributes a useful cross check on the data from sustaining areas and hollow ways from within the same area (Wilkinson, et al. n.d.).

This model of settlement and its surrounding territory, although drawing upon a wide range of data sources, immediately highlights the problem of such linear reconstructions. Even when buttressed by cross checks, each stage in the calculation is predicated upon a number of assumptions which may be only approximately correct or correct only for certain circumstances. As the number of variables increases (such as on-site population density, crop yield per unit area, food consumption per individual, plough teams, etc.), the range of variation of the

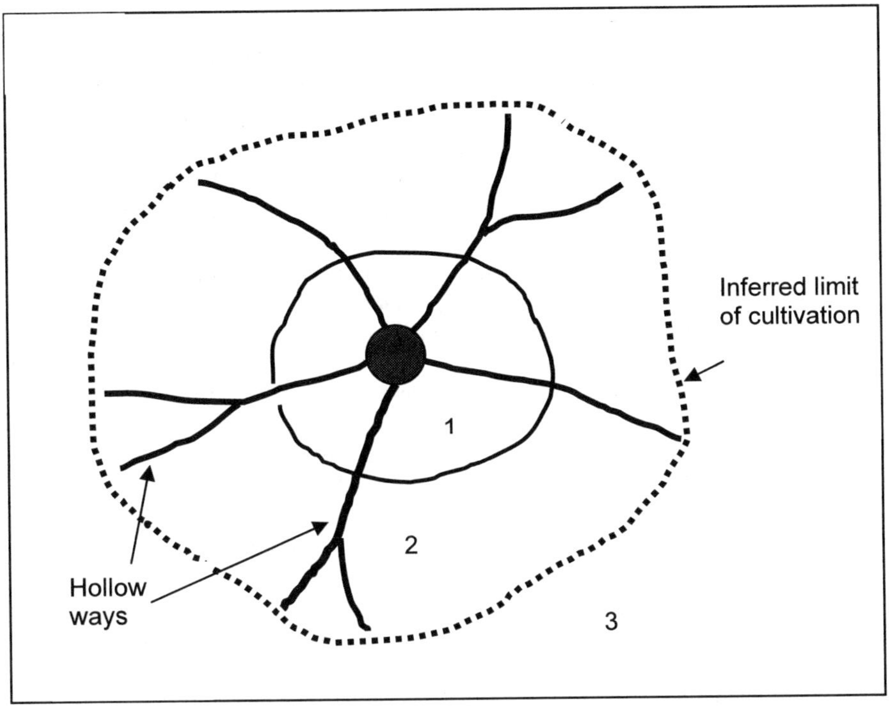

Figure 7.3 Sketch depicting the basic elements of site territory for the neighbourhood of a Bronze Age tell in Upper Mesopotamia. The inner zone (1) represents the dense field scatter, (2) the outer zone of cultivation, and (3) the pasture on the steppe beyond

estimates itself increases. As a result, the landscape reconstructions are less accurate, which diminishes their utility.

The alternative to this dilemma is to build up a fine grained and detailed model of the town, fields, pasture and indeed the entire economy so that each household can be seen to work according to its own labour force and capabilities, on its allocated patch according to a flexible set of rules (that can be changed to test different circumstances). This represents the logical next step based around agent based modelling.

Incorporating the Political Economy

It is evident from the forgoing that as the settlements analyzed become larger in size, not only do their agricultural economies become larger, but also they can cross the threshold from a subsistence to a political economy. Unfortunately, it is difficult to incorporate this increase in system complexity using traditional territorial analysis, especially if this is to be done in a non-deterministic way.

A variation of the Earle and D'Altroy model of staple and wealth economies (Earle 2002) provides a conceptual framework for the development of political economies in the ancient Near East. However, one must be wary of projecting such models back in time uncritically (see Schloen 2001:199-200). In Upper Mesopotamia three basic components of the economy can be recognized:

1. The flow of mobile high value goods via networks, potentially over long distances, either as trade, tribute, plunder or related process.
2. Staple economies are characterized by the production and distribution of mainly cereal or legume crops of low value, and low mobility. They require high volume storage space, and are relatively difficult to transport over long distances. Consequently, the distribution of staple goods is limited to a relatively short distance of the major settlements, although this depends upon the type of economy under consideration.
3. The pastoral economy is based around either village based flocks or the nomadic herding of larger flocks in the steppe or desert.

Of the above components, 1 is vulnerable to severance by nomadic groups or social or political disruptions, 2 to climatic fluctuations (as well as a range of social practices that influence land use), and 3 to climatic fluctuations, as well as variations in exchange with sectors 1 and 2.

Overall, the staple economy is relatively straightforward to model because it is focused close to the site, and in certain ways resembles the subsistence economy. On the other hand the flow of wealth and the pastoral economy are more difficult to deal with because they entail increasing the modelling framework to cover much larger areas (Figure 7.4). This is particularly acute in the case of trade and exchange because, if the modelling framework is to be expanded, then it is necessary to scale up the entire model, to encompass innumerable additional settlement systems. Similar problems of increasing system scale arise when dealing with pastoral economies that range over large areas of desert or steppe, although in both cases it is possible to side-step the problems by adopting certain simplifying mechanisms. Incorporating these fundamental features into the analysis of site and regional economies is a major challenge for archaeologists, but one that can be met by harnessing the analytical capabilities of agent based models.

Agent Based Models and the MASS Project

Whereas earlier techniques of site catchment analysis and sustaining area estimation could be conducted with relatively modest outlays of effort, agent based models require large inputs of research time in order to produce a fine grained

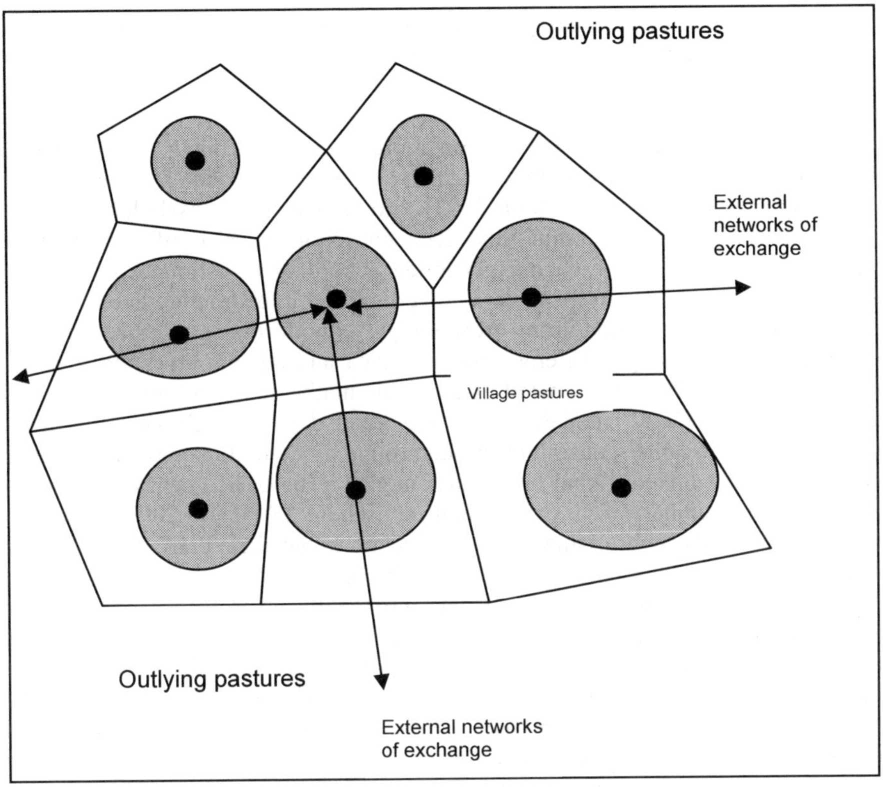

Figure 7.4 Sketch of an enlarged landscape associated with a simple political economy. Arrows indicate the direction, but not the scale, of the enlarged catchment of the wealth or network economy. Shaded areas are cultivated land, neighbouring areas in white are the village pasture lands, areas beyond are the outlying pastures

computer based system for the reconstruction of an entire agricultural and social system. In recent years agent based models have had a considerable impact on the social sciences because they enable computers to simulate the actions of heterogeneous agents that populate physical or social landscapes (Bentley 2003b:20). According to Bentley (2003b:21), unlike more traditional models (such as the territorial models described above) agent based models:

1. do not depend upon the traditional assumptions of equilibrium, normality and linearity
2. frequently demonstrate emergent phenomena

3. compared to generalizing mathematical approaches, are a more natural way to describe a social system because they replicate the actions of agents.

The present summary is based upon more detailed accounts of the MASS Project (MASS Project 2005) and pioneering work by the Argonne DIS division on social modelling (Christiansen 2000, 2003). Individuals and households form the basic agents of the MASS model, and each household consists of a number of members, the actual number being generated according to estimates of ancient demographic figures. The entire social and agricultural infrastructure is built up on the basis of evidence culled from archaeology, cuneiform texts, ethnographies, consultants' reports, long-term environmental data and other sources (Wilkinson, et al. n.d.). Crop yields are estimated by available mathematical agronomic models (such as USDA's SWAP model), which use input data from climate, estimated either by random weather generators (based around modern climate parameters) or theoretically from linked global circulation models. The basic agent employed is the individual member of a patrimonial household (Schloen 2001), and the simulation proceeds by allowing the agents to go about their daily life, growing and processing crops throughout their entire life cycle.

Cuneiform texts provide one of the most valuable data sources for our simulations because they relate (in most cases) to the period of time of the model. Unfortunately, caution is required in their interpretation because they relate to state level societies, the acts of kings or other rules, and generally under-represent the activities of everyday people (Zimansky 2005). One outcome of the use of texts in modelling small scale societies is that, ironically, the very existence of cuneiform texts at a site demonstrates already the existence of a significant degree of external influence (or perhaps even control). In other words, the site in question must have been part of a much larger political economy and that the increased information provided by the texts, in terms of quantitative modelling at least, is more than off-set by the increase in scale of the system to be analyzed. This increase in system size and complexity would require a massive increase in model scale to accommodate it. To what degree such problems seriously affect the model is still to be determined, but they act as a warning that increased information comes at a cost in terms of a necessary increase in the scale of analysis.

To date, simulations have been undertaken for model runs of up to 100 years during which time individual households can be seen to evolve and therefore fluctuate in size because their respective agents develop according to small, and slightly different, sets of initial conditions (more or less land, variations in the size of household labour pools, etc.). Consequently, subsequent generations of households will differ again, because of changing food levels, storage, exchange with neighbours or kin groups, and so on. Therefore not only does the household vary in size throughout one generation (Gallant 1991), some households will increase in size and/or affluence whereas others will decrease, or even perish and disappear from the system. Similarly household possessions, number of animals owned, area of land (or land shares or usufruct rights) and so on, will also change or evolve with time.

At their most basic, such models only reconstruct the agricultural subsistence economy of individual communities, but by scaling up the system to include neighbouring settlement and agricultural systems it will become possible to inject not only greater geographical variation, but also to introduce elements of the political economy, such as tribute, flows of bulk cereals within a staple economy and even the flow of wealth through an enlarging system of networks. However, caution must be exercised when scaling up the model universe because dynamics at one observational scale cannot necessarily be extrapolated to other scales (McGlade 2003:116). By extending the geographical area of the model to incorporate an increasing number of settlements, the scale and complexity of the model will increase accordingly.

In addition to the staple economy, it is necessary to incorporate a pastoral sector, comprising flocks of sheep and goats that must be taken to outlying pastures to graze every day. This again increases both the scale and complexity of the system being analyzed. Such pastoral usage can be modelled using existing off-the-peg land use and grazing models, but if the size of the flocks exceeds the capacity of the village pastures to accommodate the flocks, then modelling must be capable of dealing with a pastoral system that extends well out into the steppe or desert. Because many Bronze Age cities were well known to have enormous holdings of animals (Gelb 1986), such flocks and their often remote pastures must be incorporated into the modelling framework. This becomes increasingly difficult to achieve as the geographical spread of the model increases, and this subject represents a challenge for future modelling efforts.

Trade networks, which in the Near East can entail the movement of high value goods such as metals and obsidian over 100s of km, also result in increases in system scale. Simply to enlarge the spatial limits of the models to cover many 1000s of square kilometers would entail not only a massive increase in computer effort (i.e., to incorporate and model new systems of settlements), but more importantly would require the incorporation of little known communities, whose economy cannot be modelled with any degree of sensitivity because they might fall outside the limit of existing surveys. As an alternative to incorporating ever larger regional economic systems, other methods, such as network analysis must be employed (Watts 1999).

As noted above, parts of the Khabur basin of Syria supply compelling evidence for the existence of Bronze Age communication networks which take the form of networks of hollow way routes radiating from a series of primarily Bronze Age tell sites (Figure 7.5; Sallaberger and Ur 2004; Ur 2004). This network of radial routes links settlements with their fields, as well as providing cross country routes. A noteworthy feature of Figure 7.5 is that the main regional capital of the third millennium B.C. (Tell Brak = ancient Nagar) is evident not simply by virtue of its size, but also by the large number of routes radiating from it. In other words it appears to have formed a major nodal hub within a series of lesser nodes distributed throughout the region. Overall, this regional system of routeways

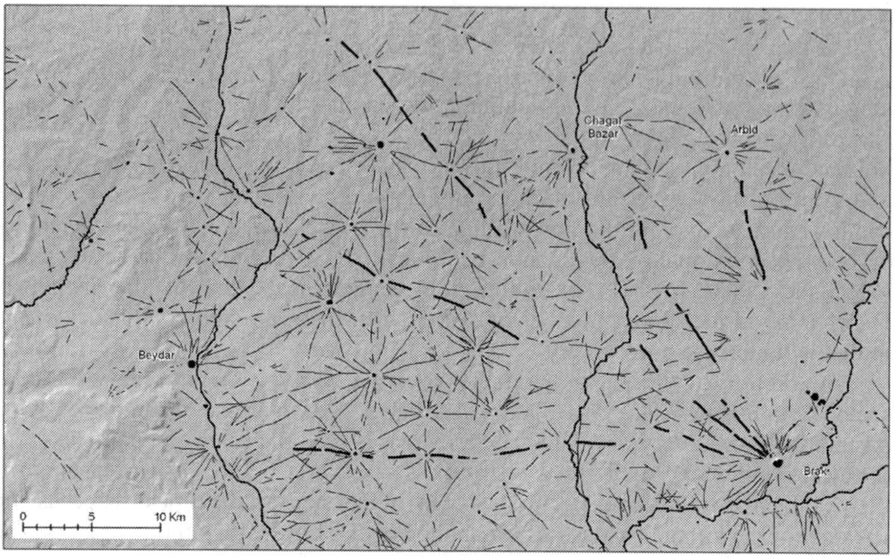

Figure 7.5 Hollow way routes mapped for the central Khabur basin to the north and west of Tell Brak, from Ur 2004

appears to have contributed to the process of fourth and third millennium urbanization by reinforcing the nodality of individual tells, which fostered further demographic growth at those centres, which in turn attracted more routes per node in a mutually reinforcing feedback process. Consequently the network, as indicated by the hollow ways routes, is not simply a passive feature of the system. It is itself an active agent of the cycle of growth.

Dynamics Versus Statics of Territorial Models

A general failure of site catchment analysis was its rather deterministic approach to the inference of paleo-economies, which could easily result in unwarranted assumptions being made about how the ancient communities obtained their livelihoods. Sustaining area analysis equally included some major assumptions which now can be improved upon considerably.

Even landscape analyses, which incorporate the greatest amount of primary archaeological data, can be improved upon. Not only are there wide error margins in the demographic estimates employed, but earlier manifestations of these models can attract accusations of being structured as closed systems (e.g., Butzer's 1997 critique of Wilkinson 1994). Although the later versions of landscape models do allow for the functioning of an external political economy, as well as some degree of exchange beyond the immediate settlement, they still suffer from the disadvantage of presenting settlement land use systems within an equilibrium framework in which the cultivated area (represented by hollow ways or field scatters) represents the 'maximum extent' of the system, and the entire system appears to have attained a state of dynamic, homeostatic equilibrium. Despite such criticisms the landscape analyses represent a significant advance over earlier models by allowing archaeologists to develop reconstructions of agricultural landscapes for certain idealized snapshots in time.

The introduction of agent based models injects a dynamism that is more than simply novel. Such simulations must be used with care however, especially in the case of what might be described as 'full system modelling', in which the economic systems under study are captured in as much detail as possible. In such models there is a danger of adding too much detail which may make the desired results of the simulation almost inevitable (Bentley 2003b:22, citing Inchiosa and Parker 2002). Nevertheless such simulations are useful by allowing simulations to develop simply as a result of the operation of internal interactions without the impact of external stresses.

On the other hand, the MASS Project has undertaken various simulations that essentially function as laboratory experiments. For example, such 'experiments' tackle questions such as 'how would the demographic system respond if the community were impacted by a run of (say) five years of drought?', or 'what would happen if 90% of the adult male population were withdrawn from the simulation for a period of 6 months from March to September (perhaps as corvée labour or to fight a war)?'. In the case of the labour withdrawal scenario one can see that the agents responded by initiating a flurry of exchange of animals for grain to make up for the deficit in grain harvested and processed. As a result there was a spike of animal exchanges recorded in the output data for the year of labour withdrawal (Figure 7.6). Such exchanges result in a flow of animals (or debt) to certain households, thereby introducing an element of history into the simulations.

In the case of the MASS simulations, the households themselves can be seen to evolve, develop or decline through the simulations (currently for the MASS project attaining a 100 year maximum duration). Whereas some families grow in size and resources, others become impoverished so that their members, if they are to survive at all, might become attached to more successful families. Consequently one can anticipate the development of larger 'elite' households alongside client or other dependent households. Clearly, for such processes to operate within the model it is necessary to incorporate the appropriate mechanisms that will allow such things to happen, according to a defined protocol, but even in the current generation of century length models we can see some degree of development along these lines. A crucial outcome of agent based models of settlement and land use systems therefore is that there appears to be some degree of social development or evolution within them. Rather than being imposed, this is a natural emergent outcome of the model itself.

Moreover, with the development of wealthy or subordinate households, a subsistence economy may be expected to evolve into a more complex political

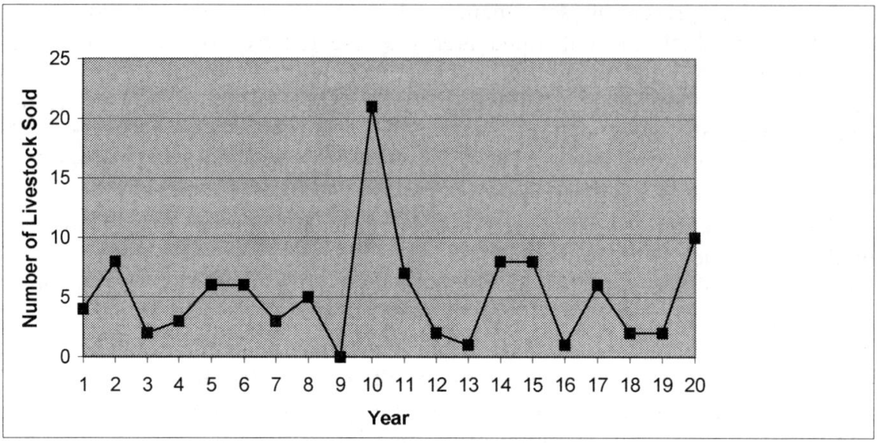

Figure 7.6 Result of the withdrawal of labour from a model community as corvée or to fight a war. The spike shows the increased number of exchanges of animals (sold) for grain (purchased). Simulation by J. Christiansen, courtesy of ANL Division of Information Sciences

economy comprising different levels of inter-dependency. The incorporation of mechanisms that allow for the exchange of goods (Reynolds and Kobti 2003) not only safeguards the household economy during episodes of food stress, it also generates a vigorous network of exchange that feasibly might be incorporated into the developing regional exchange network. We have yet to incorporate palaces or kings into our models primarily because these would be expected to develop, not by imposition, but as an emergent property of the patrimonial household itself.

An additional deficiency of traditional territorial models was their inability to handle the dynamic relationship between settlement and territory as well as different sectors of the political economy. Agent based models can at least lead to the erosion of such barriers. For example, modelling by the MASS group suggests that if an event such as a run of dry years (5 years in this case) occurs then the adaptive response of individual agents (or household members) is to a) increase the area of land under cultivation, b) increase the exchange of staple goods for other items such as animals, or c) incur debt or exchange obligations. In other words, a climate event in the form of a drought, translates into a socio-economic process such as exchange or increased obligations to kin group.

Significantly, the effect of processes such as exchange of animals for grain is well attested in the ethnographic and geographical literature which suggests that the agents are following a course of action that is replicated in the real world (Mortimore 1989). The cumulative effect of such transactions over many generations may therefore result in certain households or kin groups accumulating large holdings of sheep or goats, which would exceed the capacity of the local village pastures. Such processes might account for the enormous, and historically attested, flocks of sheep and goats known to have been attached to Bronze Age centres such as Ebla (Gelb 1986).

Climatic factors (or labour shortages) that contribute stresses to the staple economy, could be transmitted into local exchange systems and perhaps ultimately into the exchange components of the wealth economy and their networks. At present such translations from sector to sector are hypothetical, because model durations are, at present, limited to century-scale runs which are too short to demonstrate such evolutionary developments. Nevertheless, they emphasize the utility of using agent based models for modelling territorial settlement relationships.

Although the MASS simulations remain at a fairly rudimentary level, they exhibit a potential for allowing small scale subsistence communities to develop into political economies of larger scale. In order to develop into even larger or more expansive state level societies, mechanisms for aggregation of small-scale polities, domination, conflict, or subordination will be required. How the tension between 'bottom up' and 'top-down' processes will manifest itself is, as yet, unclear.

By employing the patrimonial household as the basis of the model framework, letting these households develop through time within a flexible political economy, suitably networked as implied by the landscape data, it is anticipated that archaeologists will be able to develop a completely new suite of models which will enable some level of human agency to operate. Such models will be dynamic and allow modelled communities to evolve, or transform from say sedentary agrarian to pastoral communities and also, ultimately, to collapse.

To conclude, the approach to modelling employed in this chapter clearly represents a development of processual thought. However, by incorporating human agency as an active part of the analysis it is now possible to counteract some valid criticisms of processual models. The models sketched herein will not necessarily satisfy many of a post-processual persuasion because what has been described as 'landscape as experienced' (Preucel and Hodder 1996:33) as well as symbolic and cognitive elements of social behaviour remain to be fully developed. Nevertheless, just as traditional deterministic processes of GIS have been adapted by archaeologists to contribute to human cognition of landscapes (e.g., viewshed analysis), it is hoped that the variety of approaches offered by Complex Adaptive Systems will become adapted so that a wider and more creative range of models can be developed. By building the individual, the household, and kin groups actively into the analysis of settlements and their economies, these models are already incorporating aspects of more recent archaeological thought. Moreover, because evolutionary processes are evident in our models, Complex Adaptive Systems also demonstrate the potential to contribute to the development of evolutionary archaeology.

Acknowledgements – Much of this chapter draws from the project 'Modelling Bronze Age Settlement Systems in a Dynamic Environment' (MASS) which is a collaborative effort of the Oriental Institute and Department of Anthropology at the University of Chicago, the University of Edinburgh, and Argonne National Laboratory. I am particularly grateful to the National Science Foundation Program: Dynamics of Coupled Natural and Human Systems (Grant No. 0216548), for providing major funding for the MASS project. Earlier exploratory work was funded by the University of Chicago/Argonne National Laboratory Collaborative Grants Program in 1998 and 1999 as well as by the University of Chicago, Advanced Technology Initiative (2000). I must also thank member of the MASS group, namely McGuire Gibson, John Christiansen, Magnus Widell, David Schloen, Nick Kouchoukos, Chris Woods, John Sanders, Kathy-Lee Simunich, Mark Altaweel, Jason Ur, Carrie Hritz, Scott Branting, Jake Lauinger, Tate Paulette, and Jonathan Tenney for animated discussion during numerous seminars of some of the issues examined in this chapter. Certain of the ideas presented here have been discussed during these seminars, although the group members may not agree with all of the ideas expressed.

Chapter 8

Afterword

J. Stephen Lansing
University of Arizona & Santa Fe Institute

Robert L. Axtell
The Brookings Institution

Nonlinear dynamics or 'complexity' models entered the discourse of archaeology rather suddenly in the 1990s, with the appearance of two sophisticated agent based models of prehistoric societies in the American Southwest. These models came as a revelation to many archaeologists. But while some were quick to seize the tools and begin planning simulation experiments of their own, others reacted by taking a step back in order to assess the broader implications of our newfound ability to simulate complex nonlinear patterns of change. Most of the contributors to this volume belong to the latter group.

This bifurcation into theorists and hands-on modellers is not unique to archaeology; indeed we come rather late to the party. James Gleick's wonderful popular book *Chaos: Making a New Science* (1987) introduced many researchers to the mathematics of nonlinear dynamics. But while Gleick's history of the discovery of chaos was a great story, his central theme – the ubiquity of chaos – had disquieting implications. Gleick quotes mathematician Stanislaw Ulam's amusing aphorism: 'to speak of "nonlinear science" is like calling zoology the study of "nonelephant animals."' But if nonlinear systems are not only common but also generally chaotic, then researchers interested in prediction might as well put down their tools and head for the beach. Perhaps the ultimate lesson of the historical sciences like paleontology, archaeology or ecology would be the importance of random events.

But not long after *Chaos* became a bestseller, a new spate of articles on nonlinear systems began to appear in journals like *Physica D* heralding the discovery of a different phenomenon, informally known as 'anti-chaos' or 'complexity'. The precise meaning of these terms was a little mysterious, but they arrived on the scene accompanied by mathematical models and simulations which showed how seemingly random events at one level could produce orderly patterns at another. In 1993, Oxford published a book by biologist Stuart Kauffman entitled *The Origins of Order: Self-Organization and Selection in Evolution.* Kauffmann took his readers on a 709 page tour of mathematical and computational models illustrating the concept of spontaneous 'order for free', and asking why biological systems vary in their inherent capacity to adapt. The book was lauded by such distinguished scientists as John Maynard Smith, Richard Lewontin, Stephen Jay Gould, Manfred Eigen, and Philip Anderson, and advance copies were passed around by excited graduate students.

Meanwhile, anti-chaos was also being discovered by computer scientists and game theorists. One of the most intriguing examples emerged from a synthesis of these fields: John Holland's genetic algorithms. These programs did not simply mimic evolutionary processes, instead they implemented natural selection *in silico*, turning random snippets of computer code into coherent problem-solving programs. Models such as Kauffmann's NK Boolean nets and Holland's genetic algorithms provided compelling illustrations of mechanisms by which nonlinear dynamical systems could avoid chaos and, indeed, generate spontaneous order. For example, tropical ecologist Tom Ray used Holland's genetic algorithms to create an artificial world of self-replicating digital organisms called *Tierra*, in which each individual competes for memory space in order to propagate its genes within a computer. Mutation and digital sex (recombination of strings of code) enabled an evolutionary process to occur, which produced not only winners and losers, but complex ecologies replete with phenomena resembling parasitism, predator-prey dynamics, and symbiosis. It was not long before economists were also experimenting with digital organisms to simulate competition among stock traders, hoping to discover whether complex patterns could emerge, as in *Tierra*, from competition among populations of 'agents' constructed in computer code.

The idea that similar principles might be involved in the self-organization of ecologies, stock markets or genes led to the bifurcation of research interests that we now see emerging among archaeologists. On the one hand we have the modellers, and on the other the theorists who focus on the implications of the models and the underlying mathematics. Sometimes the two are one: many modellers are also theoreticians. As Christopher Beekman and William Baden observe in the introduction to this book, the interdisciplinary Santa Fe Institute came into existence in 1985, when it was becoming apparent that models of complex systems often yielded insights that could be relevant to many fields, not just the discipline in which they were originally conceived.

But this is not the place to review the history of 'complexity science'. There is already quite a literature on this topic, ranging from scientific journals to popularizations that pick up where Gleick left off.[1] Instead, I wish to consider some of the themes and questions specific to archaeology that are raised by the contributors to this book.

In their introduction Beekman and Baden ask whether complexity is better understood by archaeologists as a 'cluster of methods capable of being used by different theoretical schools', or a consistent theoretical 'package'. For Beekman,

---

[1] Some recent overviews are: Bentley and Maschner (2003), Lansing (2002, 2003), Levin (2003), and Bentley and Maschner (2003). My favourite popularization, now somewhat out of date, is Waldrop (1992).

this issue turns on the question of agency, which he sees as the central problem for archaeological theory. Beekman argues that nonlinear dynamics offers a better way to think about agency in an archaeological context. Hommon agrees, and comments favourably on the capacity of nonlinear models to specify discrete 'schemas' of rules followed by agents, and observe their effects on collective behaviour. Crumley observes that such models are well suited to exploring the behaviour of heterarchies, and focuses on the significance of information flow in the adaptation of heterarchical societies. And several authors pick up the critical issue of scale, which has been a central theme in ecology for the past two decades. How do the actions of individuals on short time scales in their immediate environments generate patterns at larger scales of space and time? The argument echoes that of Simon Levin in one of the most-cited papers in ecology:

> The key to prediction and understanding lies in the elucidation of mechanisms underlying observed patterns. Typically, these mechanisms operate at different scales than those on which the patterns are observed; in some cases, the patterns must be understood as emerging from the collective behaviours of large ensembles of smaller scale units. In other cases, the pattern is imposed by larger scale restraints (Levin 1992: 1943).

Beekman and Baden suggest that the contributors to this book share a coherent view of the attractions of the 'complexity' perspective. That is, while the authors address different topics they express a common enthusiasm for the ability of nonlinear models to capture the ways in which local interactions in heterogeneous populations can produce higher-level patterns of order, in the way that Levin suggests. The chapters by Beekman and Baden bracket the two poles of this continuum: Beekman is interested in the agency and volition of individuals, while Baden searches for global patterns of change. I wish to push this view a step further, by suggesting that nonlinear dynamical models offer archaeologists a more direct relationship between data and theories. The argument goes like this:

Archaeological data are usually treated by organizing them into taxonomic classifications that unfold in time as what Michael Barton calls 'chronosequences'.[2] Nearly everything that comes out of an excavation is docketed and categorized. Chronosequences are organized by the principle of sameness: as long as the artefact assemblages stay more or less the same, they belong in the same period. When things become different, or in other words when the passage of time reveals not more of the same but discontinuous change, there is a need for explanation. Discontinuous change is synonymous with nonlinearity. The goal of theory is to explain patterns of change in chronosequences. At a higher level, one would like to know whether similar causes are at work at different archaeological

[2] Barton (*personal communication*) defines chronosequences as 'artificial' (i.e., investigator imposed) snapshots of a dynamic past. Such snapshots are imposed by the structure and organization of the archaeological record. They can also be thought of as samples (i.e., time slices of varying 'thickness', depending on the resolution of the data) of temporal continuity.

sites. This question can be posed for regions, like the prehistoric American Southwest or the lowland Maya. At an even higher level of abstraction, one can ask whether general principles of cultural evolution exist. The problem of taxonomic classification arises at every level in this hierarchy, from the classification of potsherds at a site to the classification of whole societies by some metric of sameness.

In principle, each discontinuous change in a chronosequence requires an explanation. Explanations minimally require two steps. First, a taxonomic classification needs to be imposed on the data, so as to distinguish periods of 'sameness' from periods of change. Next, causal mechanisms must be proposed to explain these transitions. The relationship between the actual data, the painstakingly assembled chronosequences, and what needs to be explained obviously depends critically on the granularity and scope of the taxonomy. If it appears that not much has changed over a given period, then there is little need for explanation.

The first nonlinear simulation models in archaeology introduced a different way of relating data to theory, one that is less dependent on taxonomic classification. This new approach came about in the following way. In 1990 Carla van West completed her doctoral dissertation on Anasazi sites in a region of Colorado for which superb fine-scale paleoproductivity data existed. Soon thereafter she began to work with the chair of her doctoral committee, Tim Kohler, and computer scientist Christopher Langton on an agent based model. The goal of the model was to explore relationships between the productivity of the landscape for Anasazi households, and the temporal patterns of human settlements. At about the same time, George Gumerman and Jeff Dean began to work with two modellers from the Brookings Institution, Robert Axtell and Joshua Epstein, on a similar model for a smaller Anasazi region in northeastern Arizona (Long House Valley or LHV). The two teams stayed in close touch, and the models they created are quite similar. By 1995, the Kohler-van West-Langton team was publishing the results of their initial simulations (Kohler 1995; Kohler, et al. 1995).[3]

For both teams, the goal was to explain patterns of change in chronosequences by modelling mechanisms or drivers for the features of interest. Most of the mechanisms they tested were simple linear models: so much rain produces so much maize in a given location; households grow and spread as a result of demographic change and the relative productivity of different sites on the landscape. The fewer and simpler the assumptions for each of these mechanisms, the easier it was to test their predictions. Nonlinear effects can emerge as a consequence of the interaction of the mechanisms: for example, a bottleneck caused by the weather in one region can trigger unpredictable effects elsewhere.

---

3 For an overview of both Anasazi models, see Gumerman and Kohler (2001). Also recommended is Kohler and Gumerman (2000).

The first round of simulations for the LHV model predicted prehistoric populations six times larger than those inferred from the chronosequence data. The modellers quickly realized that their crop model was based on the yield of modern maize, not the varieties available to the Anasazi. In addition, the age at which young women could get pregnant was calculated incorrectly. When these rates were recalculated, the model showed a startlingly close correlation to the archaeological data. This sort of agreement encourages us to see the potential of formal mathematical models. But it also illustrates a different point, the one I wish to emphasize here. In this case, the relationship between the data and the 'theory' (the drivers encoded in the simulation to generate predictions) does not depend on the taxonomic classification imposed on the chronosequence. Instead, the entire chronosequence is modelled, and the theory attempts to predict the whole sequence. Both teams used the same method to assess the adequacy of their theoretical predictions; time-series graphs relating predicted and observed values through time for the features of interest. This method for validation has become standard for agent based models, particularly in archaeology.

My point is that this approach tightens up the relationship between archaeological data (chronosequences) and theory (candidate explanations for patterns of change). Hitherto, I would argue that most theories addressed only a small subset of the chronosequence data, which had to be filtered through the taxonomic classification process. Explanations typically considered only a few variables, and of course lacked mathematical tools to handle nonlinear effects. The two Anasazi models demonstrated predictive power that is simply out of reach for conventional archaeological explanations.

A second notable innovation in the Anasazi simulations was to introduce a novel way to handle the question of agency – a point that is explicitly explored by Beekman in this volume, and also taken up by several other contributors. Agency arguably became the central theoretical problem for anthropology when structuralism, which does away with agency altogether, became the dominant theoretical paradigm. The demise of the unitary subject, as the late Valerio Valeri wrote, 'created paradoxes that threatened to make anthropology theoretically indefensible' (Valeri 2001:376). Post-structuralist anthropology, which goes by the name of post-processualism in archaeology, has struggled to find a niche for the subject as agent, even if only in the interpretation of the data. But for nonlinear modellers, it proved impossible to do away entirely with agency. All archaeological models include a human component, with social actors who respond to one another and to their environments. Without agency – the ability of the actors to obtain information and act on it – these models have nothing to do. In other words, while structuralist models eliminate agency, the agent-based Anasazi models require it. The question becomes how to model agency. How much information do actors need? In some more recent formulations of the 'Village' model, actors engage with a 'belief space' that exists apart from their own individual 'cognition'. The criterion for modelling agency is not the richness of the cognitive models that agents can be endowed with, but rather what kind of agency is needed to understand the patterns of change exhibited in the chronosequences.

This question is at the forefront of contemporary modelling projects in archaeology.

The issue of agency also relates to another shared characteristic of nonlinear models in archaeology, including not only the Anasazi models but also their recent imitators. That is an effort to expand the scope of the data assembled into chronosequences. As models of social actors and households become more detailed and realistic, there is a need to improve the resolution of the changing environments that they inhabit. Consequently, the teams involved in these projects are expanding to include experts in geospatial modelling, hydrology, paleoecology, climate, soil science and other fields. One such team is working on a model that will be used to carry out a comparative analysis of two regions located at opposite ends of the Mediterranean basin, in eastern Spain and the southern Levant. Such comparative projects offer the potential to test more general models of change in chronosequences. What causes sequences to diverge? Are similar factors involved in different cases? This kind of analysis will provoke new questions, some of which have been anticipated by researchers studying the question of robustness or resilience. This question has to do with the ways in which complex systems respond to perturbations. At this level, some of the mathematical theory of nonlinear dynamics will become relevant.[4] Thus one can ask, what is the basin of attraction for a given system? Agent-based models make it possible to simulate responses to perturbations, and so begin to address the fundamental question posed by Stephen Jay Gould: what if the tape were run again?

This volume describes our current understanding of the nexus of nonlinear science, complexity, agent based modelling and archaeology. But instead of viewing these chapters as a summary of a mature field, it is more plausible to suggest that they are simply the opening foray into new, uncharted terrain, in which human agency interacts with natural environment and out of which cultural institutions arise, grow and evolve. Today we have high level theories about how early humans formed the social organizations that served them successfully for millennia, how distinct social formations competed for scarce resources, and how hierarchies and heterarchies of social organizations have come into existence. But it is also true that today we have essentially no working explanatory models for any of these phenomena; for the birth of human social organizations, the emergence of complex self-governance arrangements, or patterns of resource exploitation that may or may not be sustainable and which seem inexorably to lead to cycles of societal growth and collapse. Tools derived from nonlinear mathematics like those described in this book will facilitate attempts to build models to explain such phenomena. And if the results of extant model-building efforts are any guide, we will discover that existing high level theories are incomplete: too vague and too mechanical (contra emergent) to provide the kinds of specifications needed by modellers. Filling in this gap – in essence, spanning the

---

4 For an overview of robustness, see Jen (2005). On resilience, see publications of the resilience alliance at http://www.resalliance.org/ev_en.php.

space between current social theory and archaeologically-significant chronosequences – is precisely what models can do and, arguably, is a pressing task for the continued scientific progress of archaeology.

Bibliography

9/11 Commission. (2004). *The 9/11 Commission Report: Final Report of the National Commission on Terrorist Attacks Upon the United States.* (Authorized Edition), W. W. Norton & Co., New York.

Adair, J. (1930). *Adair's History of the American Indians*, Watauga Press, Johnson City.

Adams, R. M. (1981). *Heartland of Cities*, University of Chicago Press, Chicago.

Adams, R. M. (2001), 'Complexity in Archaic States', *Journal of Anthropological Archaeology*, vol. 20, pp. 345-360.

Adams, R. N. (1975). *Energy and Structure: A Theory of Social Power*, University of Texas Press, Austin.

Adams, R. N. (1981), 'Natural Selection, Energetics, and "Cultural Materialism"', *Current Anthropology*, vol. 22, no. 6, pp. 603-624.

Adams, R. N. (1982a). The Emergence of Hierarchical Social Structure: The Case of Late Victorian England, in W. C. Schieve and P. M. Allen (eds.), *Self-Organization and Dissipative Structures*, University of Texas Press, Austin, pp. 116-131.

Adams, R. N. (1982b). *Paradoxical Harvest: Energy and Explanation in British History, 1870-1914*, Cambridge University Press, Cambridge.

Adams, R. N. (1988). *The Eighth Day: Social Evolution as the Self-Organization of Energy*, University of Texas Press, Austin.

Alam, S. J. (2004). *Ramifications of Gift Exchange in a Multiagent System.* Unpublished thesis, Department of Computer Science, Saarland University, Germany.

Albert, A. (ed.). (1995). *Chaos and Society*, IOS Press, Amsterdam.

Albrecht, W. A. (2000) 'The Drought Myth: An Absence of Water Is Not the Problem', *Acres*, vol. 30, no. 11.

Allen, P. M. (1982). The Genesis of Structure in Social Systems: The Paradigm of Self-Organization, in C. Renfrew, M. J. Rowlands and B. A. Segraves (eds.), *Theory and Explanation in Archaeology: The Southampton Conference*, Academic Press, New York, pp. 347-374.

Allen, P. M. (1989). Modelling Innovation and Change, in S. van der Leeuw and R. Torrence (eds.), *What's New? A Closer Look at the Process of Innovation*, Unwin Hyman, London, pp. 258-280.

Allen, P. M. (1994). Evolutionary Complex Systems: Models of Technological Change, in L. Leydesorff and P. van den Besselaar (eds.), *Evolutionary Economics and Chaos Theory, New Directions in Technological Studies*, St. Martin's Press, New York, pp. 1-18.

Allen, P. M. (1997a). *Cities and Regions as Self-Organizing Systems, Models of Complexity*, Gordon and Breach, Amsterdam.

Allen, P. M. (1997b). Models of Creativity: Towards a New Science of History, in S. E. van der Leeuw and J. McGlade (eds.), *Time, Process, and Structured Transformation in Archaeology*, Routledge, London, pp. 39-56.

Allen, P. M. (2001). The Dynamics of Knowledge and Ignorance: Learning the New System Science, in M. Matthies, H. Malchow and J. Kriz (eds.), *Integrative Systems Approaches to Natural and Social Dynamics*, Springer, Berlin, pp. 3-30.

Allen, P. M., and McGlade, J. M. (1987), 'Modelling Complex Human Systems: A Fisheries Example', *European Journal of Operational Research*, vol. 30, pp. 147-167.

Anderson, D. G., Stahle, D. W., and Cleaveland, M. K. (1995), 'Paleoclimate and the Potential Food Reserves of Mississippian Societies: A Case Study from the Savannah River Valley', *American Antiquity*, vol. 60, no. 2, pp. 258-286.

Archer, M. (1982), 'Morphogenesis vs. Structuration: On Combining Structure and Action', *The British Journal of Sociology*, vol. 33, pp. 455-483.

Archer, M. (1995). *Realist Social Theory: The Morphogenetic Approach*, Cambridge University Press, Cambridge.

Archer, M. (1996). *Culture and Agency: The Place of Culture in Social Theory*, 2nd edition. Cambridge University Press, Cambridge.

Archer, M. (2000). *Being Human: The Problem of Agency*, Cambridge University Press, Cambridge.

Archer, M. (2003). *Structure, Agency, and the Internal Conversation*, Cambridge University Press, Cambridge.

Archer, M., Bhaskar, R., Collier, A., Lawson, T., and Norrie, A. (eds.). (1998). *Critical Realism: Essential Readings*, Routledge, London.

Archer, M., Collier, A., and Parpoe, D. V. (eds.). (2004). *Transcendence: Critical Realism and God*, Routledge, London.

Archer, M., and Tritter, J. (eds.). (2000). *Rational Choice Theory: Resisting Colonization*, Routledge, London.

Arianoutsou, M., and Margaris, N. S. (1981), 'Fire-Induced Nutrient Losses in a Phryganic (East Meditteranean) Ecosystem', *International Journal of Biometeorology*, vol. 25, pp. 341-347.

Arnon, N. S., and Hill, W. W. (1979). Santa Clara Pueblo, in A. Ortiz (ed.), *Handbook of North American Indians, Vol. 9, Southwest*, Smithsonian Institution, Washington, DC, pp. 296-307.

Arquilla, J., and Ronfeldt, D. (eds.). (2001). *Networks and Netwars: The Future of Terror, Crime, and Militancy*, RAND, Santa Monica.

Arthur, B., and Arrow, K. (1994). *Increasing Returns and Path Dependence in the Economy*, University of Michigan Press, Ann Arbor.

Artigiani, R. (1989). A Model of Societal Self-Organization, in G. P. Scott (ed.), *Time, Rhythms, and Chaos in the New Dialogue with Nature*, Iowa State University Press, Ames, pp. 101-116.

Axelrod, R. (1984). *The Evolution of Cooperation*, Basic Books, New York.

Baden, W. W. (1987). *A Dynamic Model of Stability and Change in Mississippian Agricultural Systems.* Unpublished Ph.D. dissertation, Department of Anthropology, University of Tennessee, Knoxville.

Baden, W. W. (1995). *The Impact of Fluctuating Agricultural Potential on Coosa's Sociopolitical and Settlement Systems.* Paper presented at the 52nd Annual Southeastern Archaeological Conference, Knoxville.

Baden, W. W. (2002). *Quantifying Potentials within Dissipative Processes: An Evaluation of Mississippian Agriculture.* Paper presented at the 101st Annual Meeting of the American Anthropological Association, New Orleans.

Baden, W. W., and Beekman, C. S. (2001), 'Culture and Agriculture: A Comment on Sissel Schroeder, Maize Productivity in the Eastern Woodlands and Great Plains of North America', *American Antiquity*, vol. 66, no. 3, pp. 505-516.

Baert, P., and de Schampheleire, J. (1988), 'Autopoiesis, Self-Organization, and Symbolic Interactionism: Some Convergences', *Kybernetes*, vol. 17, pp. 60-69.

Bailey, F. G. (ed.). (1971). *Gifts and Poison: The Politics of Reputation*, Basil Blackwell, Oxford.

Bak, P. (1994). Self-Organized Criticality: A Holistic View of Nature, in G. Cowan, D. Pines and D. Meltzer (eds.), *Complexity: Metaphors, Models and Reality*, Addison-Wesley, Reading, pp. 477-496.

Baker, R. (1974), 'Indian Corn and Its Culture', *Agricultural History*, vol. 48, pp. 94-97.

Bartram, W. (1853), 'Observations on the Creek and Cherokee Indians', *Transactions of the American Ethnological Society*, vol. 3, no. 1, pp. 11-81.

Bartram, W., and Van Doren, M. (1928). *The Travels of William Bartram*, Macy-Masius, New York.

Bateson, G. (1972). *Steps toward an Ecology of Mind*, Ballantine, New York.

Bedau, M. A. (1997), 'Weak Emergence', *Philosophical Perspectives 11*, vol. 31, Supplement, pp. 375-399.

Beekman, C. S. (2000), 'The Correspondence of Regional Patterns and Local Strategies in Formative to Classic Period West Mexico', *Journal of Anthropological Archaeology*, vol. 19, no. 4, pp. 385-412.

Beekman, C. S. (2003a), 'Agricultural Pole Rituals and Rulership in Late Formative Central Jalisco', *Ancient Mesoamerica*, vol. 14, no. 2, pp. 299-318.

Beekman, C. S. (2003b), 'Fruitful Symmetry: Corn and Cosmology in the Public Architecture of Late Formative and Early Classic Jalisco', *Mesoamerican Voices*, vol. 1, pp. 5-22.

Beekman, C. S. (n.d.a), 'Political Boundaries and Landscape in Ancient Complex Polities'.

Beekman, C. S. (n.d.b). Lineages and Houses in Late Formative Navajas and Early Classic Llano Grande, in C. S. Beekman and P. C. Weigand (eds.), *La Tradición Teuchitlan: Un Primer Intento a Síntesis*, Colegio de Michoacan, Zamora.

Beekman, C. S., Tyndall, G., Jennings, S., Beekman, K., and Calgaro, B. (in prep). *Final Report on the 2003 Season at Navajas, Jalisco, Mexico*, Report to be submitted to the Instituto Nacional de Antropología e Historia, México.

Bender, M. M., Baerreis, D. A., and Steventon, R. L. (1981), 'Further Light on Carbon Isotopes and Hopewell Agriculture', *American Antiquity*, vol. 46, no. 2, pp. 346-353.

Bentley, R. A. (2003a). Scale-Free Network Growth and Social Inequality, in R. A. Bentley and H. D. G. Maschner (eds.), *Complex Systems and Archaeology. Empirical and Theoretical Applications*, University of Utah Press, Salt Lake City, pp. 27-45.

Bentley, R. A. (2003b). An Introduction to Complex Systems, in R. A. Bentley and H. D. G. Maschner (eds.), *Complex Systems and Archaeology. Empirical and Theoretical Applications*, University of Utah Press, Salt Lake City, pp. 9-23.

Bentley, R. A., and Maschner, H. D. G. (2001), 'Stylistic Change as a Self-Organized Critical Phenomena: An Archaeological Study in Complexity', *Journal of Archaeological Method and Theory*, vol. 8, no. 1, pp. 35-66.

Bentley, R. A., and Maschner, H. D. G. (eds.). (2003). *Complex Systems and Archaeology: Empirical and Theoretical Applications*, University of Utah Press, Salt Lake City.

Berlin, G. L., Ambler, J. R., Hevly, R. H., and Schaber, G. G. (1977), 'Identification of a Sinagua Agricultural Field by Aerial Thermography, Soil Chemistry, Pollen/Plant Analysis, and Archaeology', *American Antiquity*, vol. 42, no. 4, pp. 588-600.

Bertuglia, C. S., Bianchi, G., and Mela, A. (eds.). (1998). *The City and Its Sciences*, Physica-Verlag, Heidelberg.

Bhaskar, R. (1975). *A Realist Theory of Science*, Leeds Books, Leeds.

Bhaskar, R. (1979). *The Possibility of Naturalism: A Philosophical Critique of Contemporary Human Science*, Harvester, Brighton.

Bhaskar, R. (1989). *Reclaiming Reality*, Verso, London.

Bhaskar, R. (1993). *Dialectic: The Pulse of Freedom*, Verso, London.

Binford, S. R., and Binford, L. R. (1968). *New Perspectives in Archaeology*, Ballantine, New York.

Bintliff, J. (1997), 'Catastrophe, Chaos and Complexity: The Death, Decay and Rebirth of Towns from Antiquity to Today', *Journal of European Archaeology*, vol. 5, no. 2, pp. 67-90.

Bintliff, J. (1999a). Structure, Contingency, Narrative and Timelessness, in J. L. Bintliff (ed.), *Structure and Contingency: Evolutionary Processes in Life and Human Society*, Leicester University Press, London, pp. 132-148.

Bintliff, J. (1999b). Settlement and Territory, in G. Barker (ed.), *Companion Encyclopedia of Archaeology (Vol. 1)*, Routledge, London, pp. 505-544.

Bintliff, J. (2000). Settlement and Territory: A Socio-Ecological Approach to the Evolution of Settlement Systems, in G. Bailey, R. Charles and N. Winder (eds.), *Human Ecodynamics*, Oxbow Books, Oxford, pp. 21-30.

Bintliff, J. (2003). Searching for Structure in the Past - or Was It "One Damn Thing after Another"? in R. A. Bentley and H. D. G. Maschner (eds.), *Complex Systems and Archaeology. Empirical and Theoretical Applications*, University of Utah Press, Salt Lake City, pp. 79-83.

Blakeslee, D. J. (2002). Fractal Archaeology: Intra-Generational Cycles and the Matter of Scale, an Example from the Central Plains, in W. A. Parkinson (ed.), *The Archaeology of Tribal Societies*, Archaeological series 15, International Monographs in Prehistory, Ann Arbor, pp. 173-199.

Blanton, R. E., Feinman, G. M., Kowalewski, S. A., and Peregrine, P. N. (1996), 'A Dual-Processual Theory for the Evolution of Mesoamerican Civilization', *Current Anthropology*, vol. 37, pp. 1-14.

Blanton, R. E., and Kowalewski, S. A. (1982). *Monte Alban's Hinterland: Part I: The Prehispanic Settlement Patterns of the Central and Southern Parts of the Valley of Oaxaca, Mexico*, Regents of the University of Michigan Museum of Anthropology, Ann Arbor.

Bloom, H. (2000). *Global Brain: The Evolution of Mass Mind from the Big Bang to the 21st Century*, John Wiley & Sons, Inc., New York.

Bluhm, R. K., and Motley, J. B. (1995). *The Soldier's Guidebook*, Brassey's Inc., Washington, DC.

Bogucki, P. (2000). How Agriculture Came to North-Central Europe, in T. D. Price (ed.), *Europe's First Farmers*, Cambridge University Press, Cambridge, pp. 197-218.

Bose, A. (1967). *A History of Anarchism*, World Press, Calcutta.

Boserup, E. (1965). *The Conditions of Agricultural Growth; the Economics of Agrarian Change under Population Pressure*, Aldine Publishing Co., Chicago.

Bourdieu, P. (1977). *Outline of a Theory of Practice*, Cambridge University Press, Cambridge.

Bourdieu, P. (1992). *The Logic of Practice*, Stanford University Press, Palo Alto.

Bowman, M. L. (1915). *Corn: Growing Judging Breeding Feeding Marketing*, Waterloo Publishing, Waterloo.

Bowman, M. L., and Crossley, B. W. (1911). *Corn: Growing Judging Breeding Feeding Marketing*, Waterloo Publishing, Waterloo, IA.

Brandt, E. A. (1994). Egalitarianism, Hierarchy, and Centralization in the Pueblos, in W. H. Wills and R. D. Leonard (eds.), *The Ancient Southwestern Community*, University of New Mexico Press, Albuquerque, pp. 9-24.

Brante, T. (2001), 'Consequences of Realism for Sociological Theory-Building', *Journal for Theory of Social Behaviour*, vol. 31, pp. 167-196.

Brettell, C. (2002), 'The Individual/Agent and Culture/Structure in the History of the Social Sciences', *Social Science History*, vol. 26, no. 3, pp. 429-445.

Bricmont, J. (1997). Science of Chaos or Chaos of Science? in P. R. Gross, N. Levitt and M. W. Lewis (eds.), *The Flight from Science and Reason*, The New York Academy of Sciences, New York, pp. 131-175.

Brown, A., Fleetwood, S., and Roberts, J. M. (eds.). (2002). *Critical Realism and Marxism*, Routledge, London.

Brown, C. (1995). *Chaos and Catastrophe Theories: Non-Linear Modelling in the Social Sciences*, Sage Publications, Thousand Oaks.

Brown, C. T., and Witschey, W. R. T. (2003), 'The Fractal Geometry of Ancient Maya Settlement', *Journal of Archaeological Science*, vol. 30, pp. 1619-1632.

Brown, C. T., Witschey, W. R. T., and Liebovitch, L. S. (2005), 'The Broken Past: Fractals in Archaeology', *Journal of Archaeological Method and Theory*, vol. 12, no. 1, pp. 37-78.

Brown, J. R. (1994). The Sanborn Field Experiment, in R. A. Leigh and A. E. Johnston (eds.), *Long-Term Experiments in Agricultural and Ecological Sciences*, CAB International, Wallingford, pp. 39-52.

Brown, W. L., and Anderson, E. (1947), 'The Northern Flint Corns', *Annals of the Missouri Botanical Garden*, vol. 34, pp. 1-29.

Brown, W. L., and Anderson, E. (1948), 'The Southern Dent Corns', *Annals of the Missouri Botanical Garden*, vol. 35, pp. 255-268.

Brumfiel, E. (1992), 'Distinguished Lecture in Archaeology: Breaking and Entering the Ecosystem - Gender, Class and Faction Steal the Show', *American Anthropologist*, vol. 94, pp. 551-567.

Brumfiel, E. (1995). Heterarchy and the Analysis of Complex Societies: Comments, in R. M. Ehrenreich, C. L. Crumley and J. E. Levy (eds.), *Heterarchy and the Analysis of Complex Societies*, (Archaeological Papers of the American Anthropological Association No. 6). American Anthropological Association, Washington, DC, pp. 125-131.

Brumfiel, E., and Fox, J. W. (eds.). (1994). *Factional Competition and Political Development in the New World*, Cambridge University, Cambridge.

Burris, V. (2001), 'The Two Faces of Capital: Corporations and Individual Capitalists as Political Actors', *American Sociological Review*, vol. 66, no. 3, pp. 361-381.

Butterwick, K. (2004). *Heritage of Power: Ancient Sculpture from West Mexico*, Metropolitan Museum of Art, New York.

Bütz, M. R. (1997). *Chaos and Complexity. Implications for Psychological Theory and Practice*, Taylor and Francis, Washington, DC.

Butzer, K. (1997). Sociopolitical Discontinuity in the Near East ca. 2200 BCE: Scenarios from Palestine and Egypt, in H. N. Dalfes, G. Kukla and H. Weiss (eds.), *Third Millennium B.C. Climate Change and Old World Collapse* (NATO ASI series 49), Springer, Berlin, pp. 245-296.

Buyanovsky, G. A., Brown, J. R., and Wagner, G. H. (1997). Sanborn Field: Effect of 100 Years of Cropping on Soil Parameters Influencing Productivity, in E. A. Paul, K. Paustian, E. T. Elliott and C. V. Cole (eds.), *Soil Organic Matter in Temperate Agrosystems*, CRC Press, New York, pp. 205-225.

Byrne, D. (1998). *Complexity Theory and the Social Sciences: An Introduction*, Routledge, London.

Cabrero García, M. T., and López Cruz, C. (1997). *Catálogo De Piezas De Las Tumbas De Tiro Del Cañón De Bolaños*, Instituto de Investigaciones Antropológicos. Universidad Nacional Autónoma de México, México, D.F.

Çambel, A. B. (1993). *Applied Chaos Theory: A Paradigm for Complexity*, Academic Press, New York.

Cameron, C. M. (1999). *Hopi Dwellings: Architectural Change at Orayvi*, University of Arizona Press, Tucson.

Campbell, B. M. S. (2000). *English Seigniorial Agriculture, 1250-1450*, Cambridge University Press, Cambridge.

Campbell, B. M. S., and Overton, M. (eds.). (1991). *Land, Labour, and Livestock: Historical Studies in European Agricultural Productivity*, Manchester University Press, Manchester.

Campbell, S. (1978), 'Restricted Access to Knowledge in Vakuta', *Canberra Anthropology*, vol. 1, no. 3, pp. 1-11.

Cancian, F. (1965). *Economics and Prestige in a Maya Community: The Religious Cargo System in Zinacantan*, Stanford University Press, Stanford.

Cancian, F. (1976), 'Social Stratification', *Annual Review of Anthropology*, vol. 5, pp. 227-248.

Carabelea, C., Boissier, O., and Florea, A. (2004). Autonomy in Multi-Agent Systems: A Classification Attempt, in M. Nickles, M. Rovatsos and G. Weiss (eds.), *Agents and Computational Autonomy. Potential, Risks, Solutions* (Lecture Notes in Artificial Intelligence 2969), Springer, Berlin, pp. 103-113.

Carley, K. M. (1997). Organizations and Constraint-Based Adaptation, in A. E. Raymond, S. Horsfall and M. E. Lee (eds.), *Chaos, Complexity, and Sociology*, Sage Publications, Thousand Oaks, pp. 229-242.

Carman, H. J., Mitchell, J., and Young, A. (1964). *American Husbandry*, Kennikat Press, Port Washington, NY.

Carroll, S. B. (2003), 'Genetics and the Making of Homo Sapiens', *Nature*, vol. 422, pp. 849-885.

Castelfranchi, C. (1998). 'Through the Minds of Agents'. *Journal of Artificial Societies and Social Simulation 1*, from http://www.soc.surrey.ac.uk/JASSS/1/1/5.html.

Castelfranchi, C., and Werner, E. (eds.). (1994). *Artificial Social Systems*, Springer-Verlag, Berlin.

Ceci, L. (1975), 'Fish Fertilizer: A Native American Practice?' *Science*, vol. 188, no. 26-30.

Chapman, R. (2003). *Archaeologies of Complexity*, Routledge, London.

Chen, R. S. (1990), 'Global Agriculture, Environment, and Hunger: Past, Present, and Future Links', *Environmental Impact Assessment Review*, vol. 10, no. 4, pp. 335-358.

Chisholm, M. (1962). *Rural Settlement and Land Use*, Hutchinson, London.

Christiansen, J. H. (2000). Facet: A Simulation Software Framework for Modelling Complex Societal Processes and Interactions, in W. F. Waite (ed.), *Proceedings of the 2000 Summer Computer Simulation Conference of the Society for Computer Simulation International, Vancouver, British Columbia, Canada, July 16-20*, Society for Computer Simulation, San Diego.

Christiansen, J. H. (2003). A Flexible Object-Based Software Framework for Modelling Complex Systems with Interacting Natural and Societal Processes, in B. O. Parks, K. M. Clarke and M. P. Crane (eds.), *Proceedings of the 4th International Conference on Integrating GIS and Environmental Modelling, Sep 2-8, 2000, Banff, Alberta, Canada*, University of Colorado, Boulder.

Clark, J. E. (2000). Towards a Better Explanation of Hereditary Inequality: A Critical Assessment of Natural and Historic Human Agents, in M-A. Dobres and J. E. Robb (eds.), *Agency in Archaeology*, Routledge, New York, pp. 92-112.

Clark, J. E., and Blake, M. (1994). The Power of Prestige: Competitive Generosity and the Emergence of Rank Societies in Lowland Mesoamerica, in E. M. Brumfiel and J. W.

Fox (eds.), *Factional Competition and Political Development in the New World*, Cambridge University, Cambridge, pp. 17-30.

Clark, N., Perez-Trejo, F., and Allen, P. M. (1995). *Evolutionary Dynamics and Sustainable Development, a Systems Approach*, Edward Elgar, Aldershot.

Clarke, D. L. (1972). *Models in Archaeology*, Methuen & Co., London.

Clarke, D. L. (1978). *Analytical Archaeology*, Columbia University Press, New York.

Coale, A. J., Demeny, P., and Vaughan, B. (1983). *Regional Model Life Tables and Stable Populations*, Academic Press, New York.

Collier, A. (1994). *Critical Realism: An Introduction to Roy Bhaskar's Philosophy*, Verso, London.

Conte, R., and Gilbert, N. (1995). Introduction: Computer Simulation for Social Theory, in N. Gilbert and R. Conte (eds.), *Artificial Societies. The Computer Simulation of Social Life*, University College of London Press, London, pp. 1-18.

Cook, T. (2002). *Nested Political Coalitions*, Praeger, New York.

Cowan, G. A., Pines, D., and Meltzer, D. (eds.). (1994). *Complexity: Metaphors, Models, and Reality*, Santa Fe Institute Studies in the Sciences of Complexity, Addison Wesley, New York.

Crick, M. R. (1982), 'Anthropology of Knowledge', *Annual Review of Anthropology*, vol. 11, pp. 287-313.

CrimethInc. Workers' Collective. (2001). *Days of Love, Nights of War: Crimethink for Beginners*, CrimethInc. Free Press, Atlanta.

Cronon, W. (1983). *Changes in the Land: Indians, Colonists, and the Ecology of New England*, Hill and Wang, New York.

Crumley, C. L. (1979). Three Locational Models: An Epistemological Assessment for Anthropology and Archaeology, in M. B. Schiffer (ed.), *Advances in Archaeological Method and Theory* (vol. 2), Academic Press, New York, pp. 141-173.

Crumley, C. L. (1987). A Dialectical Critique of Hierarchy, in T. C. Patterson and C. W. Gailey (eds.), *Power Relations and State Formation*, American Anthropological Association, Washington, DC, pp. 155-168.

Crumley, C. L. (1995). Heterarchy and the Analysis of Complex Societies, in R. M. Ehrenreich, C. L. Crumley and J. E. Levy (eds.), *Heterarchy and the Analysis of Complex Societies* (Archaeological Papers of the American Anthropological Association No. 6), American Anthropological Association, Washington, DC, pp. 1-5.

Crumley, C. L. (2001). Communication, Holism, and the Evolution of Sociopolitical Complexity, in J. Haas (ed.), *Leaders to Rulers: The Development of Political Centralization*, Plenum, New York, pp. 19-33.

Crumley, C. L. (2003). Alternative Forms of Societal Order, in V. L. Scarborough, J. Fred Valdez and N. Dunning (eds.), *Heterarchy, Political Economy, and the Ancient Maya: The Three Rivers Region of the East-Central Yucatan Peninsula*, University of Arizona Press, Tucson, pp. 136-145.

Crumley, C. L. (n.d.a). Historical Ecology: Integrated Thinking at Multiple Temporal and Spatial Scales, in *World System History and Global Environmental Change Symposium, Lund University, Sweden*, Columbia University Press, New York.

Crumley, C. L. (n.d.b). Archaeology in the New World Order: What We Can Offer the Planet, in *An Odyssey of Space: Proceedings of the 2001 Chacmool Conference*, University of Calgary Press, Calgary.

Crumley, C. L., and Marquardt, W. H. (eds.). (1987). *Regional Dynamics: Burgundian Landscapes in Historical Perspective*, Academic Press, San Diego.

Cunningham, B. (2001), 'The Reemergence of "Emergence"', *Philosophy of Science*, vol. 68, no. 3, pp. S62-S75.

Darmody, R. G., and Peck, T. R. (1997). Soil Organic Carbon Changes through Time at the University of Illinois Morrow Plots, in E. A. Paul, K. Paustian, E. T. Elliott and C. V. Cole (eds.), *Soil Organic Matter in Temperate Agrosystems*, CRC Press, New York, pp. 161-169.

Dawkins, R. (1982). *The Extended Phenotype*, Oxford University Press, Oxford.

Day, G. M. (1953), 'The Indian as an Ecological Factor in the Northeastern Forest', *Ecology*, vol. 34, no. 329-346.

De Landa, M. (2000). *A Thousand Years of Non-Linear History*, Zone Books, New York.

Dean, J. S., Gumerman, G. J., Epstein, J. M., Axtell, R. L., Swedlund, A. C., Parker, M. T., et al. (2000). Understanding Anasazi Culture Change through Agent-Based Modeling, in T. Kohler and G. Gumerman (eds.), *Dynamics in Human and Primate Societies: Agent-based Modelling of Social and Spatial Processes*, Oxford University Press, Oxford, pp. 179-206.

Denevan, W. M. (1992), 'The Pristine Myth: The Landscape of the Americas in 1492', *Annals of the Association of American Geographers*, vol. 82, pp. 369-385.

Diamond, J. (2004). *Collapse: How Societies Choose to Fail or Succeed*, Viking, New York.

Diehl, M. W. (2005), 'Morphological Observations on Recently Recovered Early Agricultural Period Maize Cob Fragments from Southern Arizona', *American Antiquity*, vol. 70, no. 2, pp. 361-375.

Dietler, M., and Herbich, I. (1998). Habitus, Techniques, Style: An Integrated Approach to the Social Understanding of Material Culture and Boundaries, in M. Stark (ed.), *The Archaeology of Social Boundaries*, Smithsonian Institution Press, Washington, DC, pp. 232-263.

Dobres, M-A., and Robb, J. E. (2000). Agency in Archaeology. Paradigm or Platitude? in D. Marcia-Anne and J. E. Robb (eds.), *Agency in Archaeology*, Routledge, New York, pp. 3-17.

Dockstader, F. (1979). Hopi History 1850-1940, in A. Ortiz (ed.), *Handbook of North American Indians, Vol. 9, Southwest*, Smithsonian Institution, Washington, DC, pp. 524-532.

Donahue, B. (2004). *The Great Meadow: Farmers and the Land in Colonial Concord*, Yale University Press, New Haven.

Doran, J., Palmer, M., Gilbert, N., and Mellars, P. (1994). The EOS Project: Modelling Upper Palaeolithic Social Change, in N. Gilbert and J. Doran (eds.), *Simulating Societies. The Computer Simulation of Social Phenomena*, University College of London Press, London, pp. 195-222.

Doran, J. E. (2000). Trajectories to Complexity in Artificial Societies: Rationality, Belief, and Emotions, in T. A. Kohler and G. J. Gumerman (eds.), *Dynamics in Human and Primate Societies: Agent Based Modelling of Social and Spatial Processes* (Santa Fe Institute Studies in the Sciences of Complexity), Oxford University Press, New York, pp. 89-105.

Dozier, E. P. (1966), 'Factionalism at Santa Clara Pueblo', *Ethnology*, vol. 5, no. 2, pp. 172-185.

Dunnell, R. C. (1989). Aspects of Application of Evolutionary Theory in Archaeology, in C. C. Lamberg-Karlovsky (ed.), *Archaeological Thought in America*, Cambridge University Press, Cambridge, pp. 35-49.

Duvick, D. N., and Cassman, K. G. (1999), 'Post-Green Revolution Trends in Yield Potential of Temperate Maize in the North-Central United States', *Crop Science*, vol. 39, no. 6, pp. 1622-1630.

Earle, T. (2002). *Bronze Age Economics: The Beginnings of Political Economies*, Westview Press, Boulder.

Edelman, S. A. (1979). *Ildefonso Pueblo*, Smithsonian Institution, Washington, DC.

Egidi, M., and Marengo, L. (1995). Division of Labour and Social Co-Ordination Modes: A Simple Simulation Model, in N. Gilbert and R. Conte (eds.), *Artificial Societies. The Computer Simulation of Social Life*, University College of London Press, London, pp. 40-58.

Ehrenreich, R. M., Crumley, C. L., and Levy, J. E. (eds.). (1995). *Heterarchy and the Analysis of Complex Societies*, Archaeological Papers of the American Anthropological Association No. 6., American Anthropological Association, Washington, DC.

Elias, N. (1969). *Psychiatry in a Changing Society*, Tavistock, London.

Elias, N. (1978). *What Is Sociology?* Hutchinson, London.

Elkin, A. P. (1964). *The Australian Aborigines*, Doubleday-Anchor, Garden City.

Ellen, R. (1982). *Environment, Subsistence, and System: The Ecology of Small-Scale Social Formations*, Cambridge University Press, Cambridge.

Ellis, F. H. (1979). Laguna Pueblo, in A. Ortiz (ed.), *Handbook of North American Indians, Vol. 9, Southwest*, Smithsonian Institution, Washington, DC, pp. 438-449.

Ellis, R. D., and Newton, N. (eds.). (2000). *The Caldron of Consciousness: Motivation, Affect and Self-Organization - An Anthology*, John Benjamins Publishing, Amsterdam.

Emerson, W. D. (1878). *History and Incidents of Indian Corn and Its Culture* (Reprint of the edition printed by Wrightson, Cincinnati), Wrightson, Cincinnati.

Emirbayer, M., and Goodwin, J. (1994), 'Network Analysis, Culture, and the Problem of Agency', *American Journal of Sociology*, vol. 99, pp. 1411-1454.

Emirbayer, M., and Mische, A. (1998), 'What Is Agency?' *American Journal of Sociology*, vol. 103, pp. 962-1023.

Enard, W., Przeworsli, M., Fisher, S. E., Lai, C. S. L., Wiebe, V., Kitano, T., et al. (2002), 'Molecular Evolution of FOXP2, a Gene Involved in Speech and Language', *Nature*, vol. 418, pp. 869-872.

Epstein, J. M., and Axtell, R. (1996). *Growing Artificial Societies: Social Science from the Bottom Up*, Brookings Institution Press, Washington, DC.

Evans, L. T., and Fischer, R. A. (1999), 'Yield Potential: Its Definition, Measurement, and Significance', *Crop Science*, vol. 39, no. 6, pp. 1544-1551.

Eve, R. A., Horsfall, S., and Lee, M. E. (1997). *Chaos, Complexity, and Sociology: Myths, Models, and Theories*, Sage Publications, Thousand Oaks.

Feinman, G., and Neitzel, J. (1984). Too Many Types: An Overview of Sedentary Prestate Societies in the Americas, in M. Schiffer (ed.), *Advances in Archaeological Method and Theory* (vol. 7), Academic Press, New York, pp. 39-102.

Fenton, W. N. (1957), 'Factionalism at Taos Pueblo, New Mexico', *Bureau of American Ethnology Bulletin*, vol. 164, pp. 297-344.

Ferris, N. (1998). "I Don't Think We Are in Kansas Anymore": the Rise of the Archaeological Consulting Industry in Ontario, in P. J. Smith and D. Mitchell (eds.), *Bringing Back the Past: Historical Perspectives on Canadian Archaeology, Mercury Series, Archaeological Survey of Canada, Paper 158*, Canadian Museum of Civilization, Ottawa, pp. 225-247.

Fischer, K., Schillo, M., and Siekmann, J. (2003). Holonic Multiagent Systems: The Foundation for the Organization of Multiagent Systems, in *Proceedings of the First International Conference on Applications of Holonic and Multiagent Systems (Holomas'03)* (Lecture Notes in Artificial Intelligence, vol. 2744), pp. 71-80.

Flannery, K. V. (1972), 'The Cultural Evolution of Civilizations', *Annual Review of Ecology and Systematics*, vol. 3, pp. 399-426.

Flannery, K. V. (1976). The Village and Its Catchment Area, in K. V. Flannery (ed.), *The Early Mesoamerican Village*, Academic Press, London, pp. 91-95.

Fligstein, N. (2001), 'Social Skill and the Theory of Fields', *Sociological Theory*, vol. 19, no. 2, pp. 105-125.

Foster, H. T. (2003), 'Dynamic Optimization of Horticulture among the Muscogee Creek Indians of the Southeastern United States', *Journal of Anthropological Archaeology*, vol. 22, pp. 411-424.

Foster, H. T. (n.d), 'A Model of Long-Term Optimal Land Use among Shifting Horticulturalists of the Southeastern United States (1715-1825)', *Human Nature*, (in review).

Freeman, W. J. (1995). Chaotic State Transitions in Brains as a Basis for the Formation of Social Groups, in A. Albert (ed.), *Chaos and Society*, IOS Press, Amsterdam, pp. 355-369.

Friedman, D. (1996). Long Term Cropping Systems, Retrieved 8/30/2001, from http://www.sare.org/htdocs/hypermail/html-home/12-html/0382.html.

Friedman, J. (1982). Catastrophe and Continuity in Social Evolution, in C. Renfrew, M. J. Rowlands and B. A. Segraves (eds.), *Theory and Explanation in Archaeology: The Southampton Conference*, Academic Press, New York, pp. 175-196.

Frith, C. D., and Frith, U. (1999), 'Interacting Minds-a Biological Basis', *Science*, vol. 286, no. 5445, pp. 1692-1695.

Gallant, T. (1991). *Risk and Survival in Ancient Greece*, Stanford University Press, Stanford.

Garcia-Mason, V. (1979). Acoma Pueblo, in A. Ortiz (ed.), *Handbook of North American Indians, Volume 9, Southwest*, Smithsonian Institution, Washington, DC, pp. 450-453.

Gardner, A. (ed.). (2004a). *Agency Uncovered. Archaeological Perspectives on Social Agency, Power, and Being Human*, University College of London Press, London.

Gardner, A. (2004b). Agency and Community in 4th Century Britain: Developing the Structurationist Project, in G. Andrew (ed.), *Agency Uncovered. Archaeological Perspectives on Social Agency, Power, and Being Human*, University College of London, London, pp. 33-49.

Gardner, F. P., Pearce, R. B., and Mitchell, R. L. (1985). *Physiology of Crop Plants* (1st), Iowa State University Press, Ames.

Geertz, C. (1980). *Negara: The Theatre State in Nineteenth Century Bali*, Princeton University Press, Princeton.

Gelb, I. J. (1986). Ebla and Lagash: Environmental Contrast, in H. Weiss (ed.), *The Origins of Cities in Dry-Farming Syria and Mesopotamia in the Third Millennium B.C.*, Four Quarters, Guilford, pp. 157-176.

Gell-Mann, M. (1992). Complexity and Complex Adaptive Systems, in J. A. Hawkins and M. Gell-Mann (eds.), *The Evolution of Human Languages* (Santa Fe Institute Studies in the Sciences of Complexity, Proceedings Volume XI), Addison-Wesley Publishing Co., Redwood City, pp. 3-18.

Gemmill, C., and Smith, C. (1985), 'A Dissipative Structure Model of Organizational Transformation', *Human Relations,* vol. 38, pp. 751-766.

Gerlach, L. P. (2001). The Structure of Social Movements: Environmental Activism and Its Opponents, in J. Arquilla and D. Ronfeldt (eds.), *Networks and Netwars: The Future of Terror, Crime, and Militancy*, RAND, Santa Monica, pp. 289-310.

Germana, J. (2001), 'Emergent Properties: A Fundamental Postulate of Classical Systems Theory in Schematic Representation', *Systems Research and Behavioral Science,* vol. 18, pp. 565-568.

Gero, J. (2000). Troubled Travels in Agency and Feminism, in M-A. Dobres and J. E. Robb (eds.), *Agency in Archaeology*, Routledge, New York, pp. 34-39.

Giddens, A. (1979). *Central Problems in Social Theory*, Macmillan, London.

Giddens, A. (1984). *The Constitution of Society: Outline of the Theory of Structuration*, University of California Press, Berkeley.

Giddens, A. (1990). Structuration Theory and Sociological Analysis, in J. Clark, C. Modgii and S. Modgii (eds.), *Anthony Giddens: Consensus and Controversy*, Falmer Press, Hants, pp. 147-164.

Giddens, A. (1991). Structuration Theory: Past, Present and Future, in C. G. A. Bryant and D. Jary (eds.), *Giddens' Theory of Structuration: A Critical Appreciation*, Routledge, London, pp. 201-221.

Gilbert, N. (1995). Emergence in Social Simulation, in N. Gilbert and R. Conte (eds.), *Artificial Societies. The Computer Simulation of Social Life*, University College of London Press, London, pp. 144-156.

Gilbert, N. (2000). Modeling Sociality: The View from Europe, in T. Kohler and G. Gumerman (eds.), *Dynamics in Human and Primate Societies: Agent-based Modelling of Social and Spatial Processes*, Oxford University Press, Oxford, pp. 355-372.

Gilbert, N., and Conte, R. (eds.). (1995). *Artificial Societies. The Computer Simulation of Social Life*, University College of London Press, London.

Gilbert, N., and Doran, J. (eds.). (1994). *Simulating Societies: The Computer Simulation of Social Phenomena*, University College of London Press, London.

Gilbert, N., and Troitzsch, K. G. (1999). *Simulation for the Social Scientist*, Open University Press, Buckingham.

Gillespie, S. (2001), 'Personhood, Agency, and Mortuary Ritual: A Case Study from the Ancient Maya', *Journal of Anthropological Archaeology*, vol. 20, pp. 73-112.

Gimblett, H. R. (2002). *Integrating Geographic Information Systems and Agent-Based Modelling Techniques for Simulating Social and Ecological Processes* (Santa Fe Institute Study in the Science of Complexity Series), Oxford University Press, Oxford.

Glance, N. S., and Huberman, B. A. (1993), 'The Outbreak of Cooperation', *Journal of Mathematical Sociology*, vol. 17, no. 4, pp. 281-302.

Glance, N. S., and Huberman, B. A. (1994), 'The Dynamics of Social Dilemmas', *Scientific American*, vol. 270, no. 3, pp. 76-81.

Glansdorff, P., and Prigogine, I. (1971). *Thermodynamic Theory of Structure, Stability and Fluctuations*, Wiley-Interscience, New York.

Gleick, J. (1987). *Chaos: Making a New Science*, Viking, New York.

Graeber, D. (2004). *Fragments of an Anarchist Anthropology*, Prickly Paradigm Press, Chicago.

Granovetter, M. (1973), 'The Strength of Weak Ties', *American Journal of Sociology*, vol. 78, no. 6, pp. 1360-1380.

Granovetter, M. (1978), 'Threshold Models of Collective Behavior', *American Journal of Sociology*, vol. 83, no. 6, pp. 1420-1443.

Granovetter, M. (1983), 'The Strength of Weak Ties: A Network Theory Revisited', *Sociological Theory*, vol. 1, pp. 201-233.

Granovetter, M. (1985), 'Economic Action and Social Structure: The Problem of Embeddedness', *American Journal of Sociology*, vol. 91, no. 3, pp. 481-510.

Green, S. W. (1980). Broadening Least-Cost Models for Expanding Agricultural Systems, in T. K. Earle and A. L. Christenson (eds.), *Modeling Change in Prehistoric Subsistence Economies*, Academic Press, New York, pp. 209-241.

Griffin, D. R. (2004). *The New Pearl Harbor: Disturbing Questions About the Bush Administration and 9/11*, Olive Branch Press, Northampton.

Gronn, P. (2004). Distribution of Leadership, in G. R. Goethals, G. J. Sorenson and J. M. Burns (eds.), *Encyclopedia of Leadership*, Thousand Oaks, Sage Publications, pp. 351-355.

Guastello, S. J. (1995). *Chaos, Catastrophe, and Human Affairs: Applications of Non-Linear Dynamics to Work, Organizations, and Social Evolution*, Lawrence Erlbaum Association Publications, Mahwah.

Gumerman, G. J., and Gell-Mann, M. (1994). *Understanding Complexity in the Prehistoric Southwest* (Santa Fe Institute Studies in the Sciences of Complexity vol. 16), Perseus Books, Philadelphia.

Gumerman, G. J., and Kohler, T. A. (1994). Archaeology and the Sciences of Complexity, in *Prehistoric Cultures as Complex Adaptive Systems*, Society for American Archaeology, Anaheim.

Gumerman, G. J., and Kohler, T. A. (2001). Creating Alternative Cultural Histories in the Prehistoric Southwest: Agent-Based Modelling in Archaeology, in D. A. Phillips, Jr. and L. Sebastian (eds.), *Examining the Course of Southwest Archaeology: The Durango Conference, September 1995*, New Mexico Archaeological Council, Albuquerque, pp. 113-124.

Gunderson, L. H., and Holling, C. S. (2002). *Panarchy: Understanding Transformations in Human and Natural Systems*, Island Press, Washington, DC.

Hall, A. D. (1905), 'The Accumulation of Fertility by Land Allowed to Run Wild', *Journal of Agricultural Science*, vol. 1, p. 241.

Hall, A. D. (1917). *The Rothamsted Experiments*, Murray, London.

Hardcastle, V. G. (2000). Dissolving Differences. How to Understand the Competing Approaches to Human Emotion, in R. D. Ellis and N. Newton (eds.), *The Caldron of Consciousness. Motivation, Affect, and Self-Organization*, John Benjamins Publishing, Amsterdam, pp. 119-131.

Harn, A. D. (1978). Mississippian Settlement Patterns in the Central Illinois River Valley, in B. Smith (ed.), *Mississippian Settlement Patterns*, Academic Press, New York, pp. 233-268.

Harvey, D., and Reed, M. (1994), 'The Evolution of Dissipative Systems', *Journal for the Theory of Social Behaviour*, vol. 17, pp. 371-411.

Harvey, D., and Reed, M. (1996). Social Science as the Study of Complex Systems, in L. D. Kiel and E. Elliott (eds.), *Chaos Theory in the Social Sciences: Foundations and Applications*, University of Michigan Press, Ann Arbor, pp. 295-323.

Hays, K. A. (1993). When Is a Symbol Archaeologically Meaningful? in N. Yoffee and A. Sherratt (eds.), *Archaeological Theory: Who Sets the Agenda*, Cambridge University Press, Cambridge, pp. 81-92.

Heidenreich, C. (1971). *Huronia: A History and Geography of the Huron Indians, 1600-1650*, McClelland and Stewart, Toronto.

Heisenberg, W. (1927), 'Über Den Anschaulichen Inhalt Der Quantentheoretischen Kinematik Und Mechanik', *Zeitschrift für Physik*, vol. 43, pp. 172-198.

Helmreich, S. (1998). *Silicon Second Nature: Culturing Artificial Life in a Digital World*, University of California Press, Berkeley.

Hendrix, P. F. (1997). Long-Term Patterns of Plant Production and Soil Carbon Dynamics in a Georgia Piedmont Agrosystem, in E. A. Paul, K. Paustian, E. T. Elliott and C. V. Cole (eds.), *Soil Organic Matter in Temperate Agrosystems*, CRC Press, New York, pp. 235-245.

Herndon, G. M. (1967), 'Indian Agriculture in the Southern Colonies', *The North Carolina Historical Review*, vol. 44, pp. 283-297.

Highfield, R. (1996). *Frontiers of Complexity: The Search for Order in a Chaotic World*, Ballantine, New York.

Hindess, B. (1986). Actors and Social Relations, in M. L. Wordell and S. D. Turner (eds.), *Social Theory in Transition*, Allen and Unwin, London, pp. 113-126.

Hodder, I. (1982). Theoretical Archaeology: A Reactionary View, in I. Hodder (ed.), *Symbolic and Structural Archaeology*, Cambridge University Press, Cambridge, pp. 1-16.

Hodder, I. (1991). Postprocessual Archaeology and the Current Debate, in R. W. Preucel (ed.), *Processual and Post-Processual Archaeologies: Multiple Ways of Knowing the Past* (Centre for Archaeological Investigations Occasional Paper No. 10), Southern Illinois University at Carbondale, Carbondale, pp. 30-41.

Hodder, I. (2000). Agency and Individuals in Long-Term Process, in M-A. Dobres and J. Robb (eds.), *Agency in Archaeology: Paradigm or Platitude?* Routledge, London, pp. 21-33.

Hodder, I., and Cessford, C. (2004), 'Daily Practice and Social Memory at Çatalhöyük', *American Antiquity*, vol. 69, no. 1, pp. 17-40.

Hodgson, G. M. (2002), 'The Evolution of Institutions: An Agenda for Future Theoretical Research', *Constitutional Political Economy*, vol. 13, pp. 111-127.

Hofstadter, D. (1979). *Gödel, Escher, Bach: An Eternal Golden Braid*, Basic Books, New York.

Holder, P. (1970). *The Hoe and the Horse on the Plains*, University of Nebraska Press, Lincoln.

Holland, J. (1992), 'Complex Adaptive Systems', *Dædalus: Proceedings of the American Academy of Arts and Sciences*, vol. 121, no. 1, pp. 17-30.

Holland, J. (1995). *Hidden Order*, Addison-Wesley Publishing Company, Reading.

Holland, J. (1998). *Emergence: From Chaos to Order*, Oxford University Press, Oxford.

Homer. (1999). *The Odyssey of Homer* (R. Lattimore, Trans.), HarperCollins Publishers, New York.

Hommon, R. J. (1976). *The Formation of Primitive States in Pre-Contact Hawaii.* Unpublished Ph. D. dissertation, Department of Anthropology, University of Arizona.

Hommon, R. J. (1986). Social Evolution in Ancient Hawaii, in P. V. Kirch (ed.), *Island Societies: Archaeological Approaches to Evolution and Transformation*, Cambridge University Press, Cambridge, pp. 55-68.

Hommon, R. J. (1996). Social Complex Adaptive Systems: Some Hawaiian Examples, in J. Davidson, G. Irwin, F. Leach, A. Pawley and D. Brown (eds.), *Oceanic Culture History: Essays in Honour of Roger Green*, New Zealand Journal of Archaeology, Dunedin, pp. 579-590.

Hommon, R. J. (2000). Can a Small Fish Swallow a Big One? Or Orwell among the Hawaiians, in M. Diehl (ed.), *Hierarchies in Action: Cui Bono* (Centre for Archaeological Investigations Occasional Paper no. 27), Southern Illinois University, Carbondale, pp. 134-151.

Hommon, R. J. (2001). The Emergence of Large-Scale Society in Hawai`i, in C. M. Stevenson, G. Lee and F. J. Morin (eds.), *Pacific 2000: Proceedings of the Fifth International Conference on Easter Island and the Pacific*, Bearsville Press, Los Osos, pp. 141-149.

Hooke, R. L. (1994), 'On the Efficacy of Humans as Geomorphic Agents', *GSA Today*, vol. 4, no. 9, pp. 217 and 224-225.

Hooke, R. L. (2000), 'On the History of Humans as Geomorphic Agents', *Geology*, vol. 28, no. 9, pp. 843-884.

Hopi Dictionary Project. (1997). *Hopi Dictionary: Hop'iikwa Lava`Ytutuveni, a Hopi-English Dictionary of the Third Mesa Dialect*, University of Arizona Press, Tucson.

Hornstein, D. T. (2005), 'Complexity Theory, Adaptation, and Administrative Law', *Duke Law Journal*, vol. 58.

Houtman, N. (2004). Humans May Surpass Other Natural Forces as Earth Movers, Retrieved August 11, 2004, from http://www.eurekalert.org/pub_releases/2004-07/uom-hms070804.php.

Humphreys, P. (1997a), 'How Properties Emerge', *Philosophy of Science*, vol. 64, no. 1, pp. 1-17.

Humphreys, P. (1997b). Emergence, Not Supervenience, in *Proceedings of the 1996 Biennial Meetings of the Philosophy of Science Association. Part II: Symposia Papers* (Philosophy of Science 64, Supplement: S337-345).

Imhoff, M. L., Bounoua, L., Ricktts, T., Loucks, C., Harriss, R., and Lawrence, W. T. (2004), 'Global Patterns in Human Consumption of Net Primary Production', *Nature*, vol. 429, pp. 870-873.

Inchiosa, M. E., and Parker, M. T. (2002), 'Overcoming Design and Development Challenges in Agent-Based Modelling Using ASCAPE', *Proceedings of the National Academy of Sciences, USA*, vol. 99, pp. 7304-7308.

Jantsch, E. (1982). From Self-Reference to Self-Transcendence: The Evolution of Self-Organization Dynamics, in W. C. Schieve and P. M. Allen (eds.), *Self-Organization and Dissipative Structures: Applications in the Physical and Social Sciences*, University of Texas Press, Austin, pp. 344-361.

Jarman, M. R., Vita-Finzi, C., and Higgs, E. S. (1972). Site Catchment Analysis in Archaeology, in P. J. Ucko, R. Tringham and G. W. Dimbleby (eds.), *Man, Settlement and Urbanism*, Duckworth, London, pp. 61-66.

Jaynes, E. T., and Rosenkrantz, R. D. (1983). *E.T. Jaynes: Papers on Probability, Statistics, and Statistical Physics*, Dordrecht, Boston.

Jen, E. (ed.). (2005). *Robust Design: A Repertoire from Biology, Ecology, and Engineering*, Oxford University Press, Oxford.

Jenkinson, D. S., Bradbury, N. J., and Coleman, K. (1994). How the Rothamsted Classical Experiments have been Used to Develop and Test Models for the Turnover of Carbon and Nitrogen in Soil, in R. A. Leigh and A. E. Johnston (eds.), *Long-Term Experiments in Agricultural and Ecological Sciences*, CAB International, Wallingford, pp. 117-138.

Johnson, C. D., Kohler, T. A., and Cowan, J. (2005), 'Modeling Historical Ecology, Thinking About Contemporary Systems', *American Anthropologist*, vol. 107, no. 1, pp. 96-107.

Johnson, G. A. (1982). Organizational Stress and Scalar Stress, in C. Renfrew, M. J. Rowlands and B. A. Seagraves (eds.), *Theory and Explanation in Archaeology: The Southampton Conference*, Academic Press, New York, pp. 389-421.

Johnson, S. (2001). *Emergence: The Connected Lives of Ants, Brains, Cities, and Software*, Scribner, New York.

Joyce, A. A. (2000). The Founding of Monte Albán: Sacred Propositions and Social Practices, in M-A. Dobres and J. Robb (eds.), *Agency in Archaeology: Paradigm or Platitude?* Routledge, London, pp. 71-91.

Joyce, A. A. (2001), 'Commoner Power: A Case Study from the Classic Period Collapse on the Oaxaca Coast', *Journal of Archaeological Method and Theory*, vol. 8, no. 4, pp. 343-385.

Joyce, A. A., and Winter, M. (1996), 'Ideology, Power, and Urban Society in Pre-Hispanic Oaxaca', *Current Anthropology*, vol. 37, no. 1, pp. 33-86.

Kamakau, S. M. (1992). *Ruling Chiefs of Hawaii* (Revised), The Kamehameha Schools Press.

Kan, M., Meighan, C., and Nicholson, H. B. (1970). *Sculpture of Ancient West Mexico: Nayarit, Jalisco, Colima*, Los Angeles County Museum of Art, Los Angeles.

Kauffman, S. A. (1993). *The Origins of Order: Self-Organization and Selection in Evolution*, Oxford University Press, New York.

Kauffman, S. A. (1995). *At Home in the Universe: The Search for Laws of Self-Organization and Complexity*, Oxford University Press, New York.

Kauffman, S. A. (1999). *Investigations*, Oxford University Press, Oxford.

Kay, R. (2001), 'Are Organizations Autopoietic? A Call for New Debate', *Systems Research and Behavioral Science*, vol. 18, pp. 461-477.

Khalil, E. L. (1995), 'Non-Linear Thermodynamics and Social Science Modelling: Fad Cycles, Cultural Development and Identificational Slips', *The American Journal of Economics and Sociology*, vol. 54, no. 4, p. 423.

Kiel, L. D., and Elliott, E. (eds.). (1996). *Chaos Theory in the Social Sciences: Foundations and Applications*, University of Michigan Press, Ann Arbor.

Kiel, L. D., and Elliott, E. (eds.). (1999). *Non-Linear Dynamics, Complexity and Public Policy*, Nova Science Publishers Inc., Commack.

Kilminster, R. (1991). Structuration Theory as a World-View, in C. G. A. Bryant and D. Jary (eds.), *Giddens' Theory of Structuration: A Critical Appreciation*, Routledge, New York, pp. 74-115.

Kim, J. (1992). "Downward Causation" in Emergentism and Nonreductive Physicalism, in A. Beckermann, H. Flohr and J. Kim (eds.), *Emergence or Reduction? Essays on the Prospects of Nonreductive Physicalism*, Walter de Gruyter, New York, pp. 119-138.

Kim, J. (1999), 'Making Sense of Emergence', *Philosophical Studies*, vol. 95, pp. 3-36.

Klein, R. G. (1999). *The Human Career* (2nd edition), University of Chicago Press, Chicago.

Knabe, T., Schillo, M., and Fischer, K. (2003). Inter-Organizational Networks as Patterns for Self-Organizing Multi-Agent Systems, in *Proceedings of the Second International Conference on Autonomous Agents and Multiagent Systems (AAMAS'03)*, pp. 1036-1037.

Kohler, T. A. (1995). *Agent-Based Modelling in Archaeology: An Overview.* Paper presented at the Workshop on Creating Artificial Anasazi Worlds, Santa Fe Institute, Santa Fe.

Kohler, T. A., and Gumerman, G. J. (eds.). (2000). *Dynamics in Human and Primate Societies: Agent-Based Modelling of Social and Spatial Processes*, Santa Fe Institute and Oxford University Press, New York.

Kohler, T. A., Kresl, J., van West, C., Carr, E., and Wilhusen, R. H. (2000). Be There Then: A Modeling Approach to Settlement Determinants and Spatial Efficiency among Late Ancestral Pueblo Populations of the Mesa Verde Region, U.S. Southwest, in T. Kohler and G. Gumerman (eds.), *Dynamics in Human and Primate Societies: Agent-Based Modelling of Social and Spatial Processes*, Oxford University Press, Oxford, pp. 145-178.

Kohler, T. A., Langton, C., and van West, C. (1995). *Swarm: An Agent-Based Simulation System and Its Application to Modelling Village Formation in Southwestern Colorado.* Paper presented at the Durango Conference on Southwest Archaeology, September 1995.

Kohler, T. A., and van West, C. R. (1996). The Calculus of Self-Interest in the Development of Cooperation: Sociopolitical Development and Risk among the Northern Anasazi, in J. A. Tainter and B. B. Tainter (eds.), *Evolving Complexity and Environmental Risk in the Prehistoric Southwest* (Santa Fe Institute Studies in the Sciences of Complexity, Proceedings Volume XXIV), pp. 169-196.

Kontopolous, K. (1993). *The Logics of Social Structure*, Oxford University Press, New York.

Kramer, C. (1980). Estimating Prehistoric Populations, an Ethnographic Approach, in M. T. Barrelet (ed.), *L'archeologie De L'iraq: Perspectives Et Limites De L'interpretation Anthropologique Des Documents*, CNRS 580, Paris, pp. 313-332.

Kuhn, T. (1970). *The Structure of Scientific Revolutions* (2nd edition), University of Chicago Press, Chicago.

Kus, S. (1981). The Context of Complexity, in S. E. van der Leeuw (ed.), *Archaeological Approaches to the Study of Complexity*, Universiteit van Amsterdam, Amsterdam, pp. 198-229.

Kuznar, L. A. (1997). *Reclaiming a Scientific Anthropology*, Alta Mira Press, Walnut Creek.

Lafitau, J. F. (1977). *Customs of the American Indians Compared with the Customs of Primitive Times* (W. N. Fenton and E. Moore, Trans. vol. III), The Champlain Society, Toronto.

Lake, M. (2000). Magical Computer Simulation of Mesolithic Foraging, in T. Kohler and G. Gumerman (eds.), *Dynamics in Human and Primate Societies: Agent-based Modelling of Social and Spatial Processes*, Oxford University Press, Oxford, pp. 107-144.

Lake, M. (2004). Being in a Simulacrum: Electronic Agency, in G. Andrew (ed.), *Agency Uncovered. Archaeological Perspectives on Social Agency, Power, and Being Human*, University College of London Press, London, pp. 191-209.

Lake, M. (2005). *The Uncertain Future of Simulating the Past.* Paper presented at the 69th Annual Meeting of the Society for American Archaeology, Salt Lake City.

Langton, C. G. (1989). *Artificial Life: The Proceedings of an Interdisciplinary Workshop on the Synthesis and Simulation of Living Systems, Held September, 1987, in Los Alamos, New Mexico*, Addison-Wesley Pub. Co., Advanced Book Program, Redwood City.

Langton, C. G. (1990), 'Computation at the Edge of Chaos: Phase Transitions and Emergent Computation', *Physica D*, vol. 42, pp. 12-37.

Langton, C. G. (ed.). (1988). *Artificial Life I* (Santa Fe Institute Studies in the Sciences of Complexity), Addison Wesley, New York.

Langton, C. G. (ed.). (1994). *Artificial Life III* (Santa Fe Institute Studies in the Sciences of Complexity), Addison Wesley, New York.

Langton, C. G., Taylor, C., Farmer, J. D., and Rasmussen, S. (eds.). (1992). *Artificial Life II*, Santa Fe Institute Studies in the Sciences of Complexity. Addison Wesley, New York.

Lansing, J. S. (1991). *Priests and Programmers: Technologies of Power in the Engineered Landscape of Bali*, Princeton University Press, Princeton.

Lansing, J. S. (2000). Anti-Chaos, Common Property, and the Emergence of Cooperation, in T. Kohler and G. Gumerman (eds.), *Dynamics in Human and Primate Societies: Agent-Based Modelling of Social and Spatial Processes*, Oxford University Press, Oxford, pp. 207-224.

Lansing, J. S. (2002), 'Artificial Societies and the Social Sciences', *Artificial Life*, vol. 8, Oct. 2002, pp. 279-292.

Lansing, J. S. (2003), 'Complex Adaptive Systems', *Annual Review of Anthropology*, vol. 32, pp. 183-204.

Lansing, J. S., and Kremer, J. N. (1993), 'Emergent Properties of Balinese Water Temple Networks: Coadaptation on a Rugged Fitness Landscape', *American Anthropologist*, vol. 95, pp. 97-114.

Larson, W. E., and Hanway, J. J. (1977). Corn Production, in G. F. Sprague (ed.), *Corn and Corn Improvement* (Agronomy No. 18), American Society of Agronomy, Madison, pp. 625-669.

Layder, D. (1985), 'Power, Structure, and Agency', *Journal for the Theory of Social Behaviour*, vol. 15, pp. 131-149.

Layder, D. (1997). *Modern Social Theory: Key Debates and New Directions*, UCL Press, London.

Le Clercq, C. (1968). *New Relation of Gaspesia* (W. F. Ganong, Trans.), The Champlain Society, Toronto.

Lehner, M. (2000). The Fractal House of Pharaoh: Ancient Egypt as a Complex Adaptive System, a Trial Formulation, in T. A. Kohler and G. J. Gumerman (eds.), *Dynamics in Human and Primate Societies: Agent-Based Modelling of Social and Spatial Processes*, Oxford University Press, Oxford, pp. 275-353.

Lemieux, P. (1995). Chaos Et Anarchie, in A. Albert (ed.), *Chaos and Society*, IOS Press, Amsterdam, pp. 211-238.

Lescarbot, M. (1968). *The History of New France* (W. L. Grant, Trans. vol. III), The Champlain Society, Toronto.

Levin, S. (1992), 'The Problem of Pattern and Scale in Ecology: The Robert H. Macarthur Award Lecture', *Ecology*, vol. 73, no. 6, pp. 1943-1967.

Levin, S. (2003), 'Complex Adaptive Systems: Exploring the Known, the Unknown and the Unknowable', *Bulletin of the American Mathematical Society*, vol. 40, no. 1, pp. 3-19.

Levy, J. E. (1992). *Orayvi Revisited, Social Stratification in an "Egalitarian" Society*, School of American Research Press, Santa Fe.

Leydesdorff, L. (2001). *A Sociological Theory of Communication: The Self-Organization of the Knowledge-Based Society*, Universal Publishers, uPUBLISH.COM.

Leydesdorff, L., and van den Besselaar, P. (eds.). (1994). *Evolutionary Economics and Chaos Theory. New Directions in Technology Studies*, St. Martin's Press, New York.

Lightfoot, K. G., Martinez, A., and Schiff, A. M. (1998), 'Daily Practice and Material Culture in Pluralistic Social Settings: An Archaeological Study of Culture Change and Persistence from Fort Ross, California', *American Antiquity*, vol. 63, no. 2, pp. 199-222.

Likens, G. E., Boorman, F., Pierce, R., and Reiners, W. (1978), 'Recovery of a Deforested Ecosystem', *Science*, vol. 199, pp. 492-495.

Lo, C. (2005). *Beyond Hierarchy: Exposing a Fulcrum for Heterarchy in Mental Models.* Unpublished Ph.D. dissertation, Department of Education and Social Work, University of Sydney, Australia.

Lorenz, E. N. (1963), 'Deterministic Nonperiodic Flows', *Journal of Atmospheric Science,* vol. 29, pp. 130-141.

Lorenz, E. N. (1993). *The Essence of Chaos*, University of Washington Press, Seattle.

Lucero, L. J. (2002), 'The Politics of Ritual: The Emergence of Classic Maya Rulers', *Current Anthropology*, vol. 44, pp. 523-558.

Ludlow, P. (2001). *Crypto Anarchy, Cyberstates, and Pirate Utopias*, MIT Press, Cambridge.

Luhmann, N. (1986). The Autopoiesis of Social Systems, in F. Geyer and J. van der Zouwen (eds.), *Sociocybernetic Paradoxes*, Sage, London, pp. 172-192.

Luhmann, N. (1995). *Social Systems*, Stanford University Press, Palo Alto.

Lyman, R. (1999, December 15). Charles Schulz, Creator of "Peanuts" Retires. *New York Times*.

Lynott, M. J., Boutton, T. W., Price, J. E., and Nelson, D. E. (1986), 'Stable Carbon Isotopic Evidence for Maize Agriculture in Southeast Missouri and Northeast Arkansas', *American Antiquity*, vol. 51, no. 1, pp. 51-65.

Macy, M. W. (1998). 'Social Order in Artificial Worlds'. *Journal of Artificial Societies and Social Simulation 1*, from http://www.soc.surrey.ac.uk/JASSS/1/1/4.html.

Mandelbrot, B. (1977). *The Fractal Geometry of Nature*, W.H. Freeman, New York.

Marcus, J., and Flannery, K. V. (1996). *Zapotec Civilization: How Urban Society Evolved in Mexico's Oaxaca Valley*, Thames and Hudson, London.

Marion, R. (1999). *The Edge of Organization: Chaos and Complexity Theories of Formal Social Systems*, Sage Publications, Thousand Oaks.

Maruyama, M. (1963), 'The Second Cybernetics: Deviation-Amplifying Mutual Causal Processes', *American Scientist*, vol. 51, no. 2, pp. 164-179.

Maschner, H. D. G. (ed.). (1996). *Darwinian Archaeologies*, Plenum Press, New York.

Maschner, H. D. G., and Bentley, R. A. (2003). The Power Law of Rank and Household on the North Pacific, in R. A. Bentley and H. D. G. Maschner (eds.), *Complex Systems and Archaeology. Empirical and Theoretical Applications*, University of Utah Press, Salt Lake City, pp. 47-60.

MASS Project. (2005). Modeling Ancient Settlement Systems (Mass), from http://oi.uchicago.edu/OI/PROJ/MASS/Mass.htm.

Maturana, H., and Varela, F. (1980). *Autopoiesis and Cognition: The Realization of the Living* (Boston Studies in the Philosophy of Science, Vol. 42), D. Reidel, Dordrecht.

May, R. M. (1976), 'Simple Mathematical Models with Very Complicated Dynamics', *Nature*, vol. 261, pp. 471-477.

Mayer, A. C. (1977). The Significance of Quasi-Groups in the Study of Complex Societies, in S. W. Schmidt, L. Guasti, C. H. Lande and J. C. Scott (eds.), *Friends, Followers and Factions*, University of California, Berkeley, pp. 47-54.

Mayntz, R. (2004), 'Mechanisms in the Analysis of Social Macro-Phenomena', *Philosophy of the Social Sciences*, vol. 34, no. 2, pp. 237-259.

Mayr, E. (2000). *What Evolution Is*, Basic Books, New York.

McCulloch, W. S. (1945), 'A Heterarchy of Values Determined by the Topology of Nervous Nets', *Bulletin of Mathematical Biophysics*, vol. 7, pp. 89-93.

McCulloch, W. S. (1989). *Embodiments of Mind*, MIT Press, Cambridge.

McGlade, J. (1990). *The Emergence of Structure: Modelling Social Transformation in Later Prehistoric Wessex*. Unpublished Ph.D. dissertation, University of Cambridge.

McGlade, J. (1997). The Limits of Social Control: Coherence and Chaos in a Prestige-Goods Economy, in S. E. van der Leeuw and J. McGlade (eds.), *Time, Process and Structured Transformation in Archaeology*, Routledge, London, pp. 298-330.

McGlade, J. (1999). Archaeology and the Evolution of Cultural Landscapes: Towards an Interdisciplinary Research Agenda, in P. J. Ucko and R. Layton (eds.), *The Archaeology and Anthropology of Landscape: Shaping Your Landscape*, Routledge, London, pp. 458-482.

McGlade, J. (2003). The Map Is Not the Territory: Complexity, Complication and Representation, in R. A. Bentley and H. D. G. Maschner (eds.), *Complex Systems and Archaeology: Empirical and Theoretical Applications*, University of Utah Press, Salt Lake City, pp. 111-119.

McGlade, J., and Allen, P. M. (1986), 'Fluctuation, Instability and Stress: Understanding the Evolution of a Swidden Horticultural System', *Science and Archaeology*, vol. 28, pp. 44-50.

McGlade, J., and van der Leeuw, S. E. (1997). Introduction: Archaeology and Non-Linear Dynamics - New Approaches to Long-Term Change, in S. E. van der Leeuw and J.

McGlade (eds.), *Time, Process and Structured Transformation in Archaeology*, Routledge, London, pp. 1-31.
McGuire, R. H., and Saitta, D. J. (1996), 'Although They Have Petty Captains, They Obey Them Badly: The Dialectics of Prehispanic Western Pueblo Social Organization', *American Antiquity*, vol. 61, no. 2, pp. 197-216.
McGuire, R. H., and Saitta, D. J. (1998), 'Dialectics, Heterarchy, and Western Pueblo Social Organization', *American Antiquity*, vol. 63, no. 2, pp. 334-336.
McIntosh, R., and McIntosh, S. K. (1999). *Beyond Chiefdoms: Pathways to Complexity in Africa*, Cambridge University Press, Cambridge.
McIntosh, R., Tainter, J. A., and McIntosh, S. K. (eds.). (2000). *The Way the Wind Blows: Climate, History, and Human Action*, Historical Ecology Series, Columbia University Press, New York.
Mead, G. H. (1934). *Mind, Self, and Society*, University of Chicago Press, Chicago.
Meggers, B. J. (1954), 'Environmental Limitation on the Development of Culture', *American Anthropologist*, vol. 56, pp. 801-824.
Meggers, B. J. (1960). The Law of Cultural Evolution as a Practical Research Tool, in G. E. Dole and R. L. Carneiro (eds.), *Essays in the Science of Culture: In Honor of Leslie A. White*, Thomas Y. Crowell, New York, pp. 302-316.
Mihata, K. (1997). The Persistence of "Emergence", in R. A. Eve, S. Horsfall and M. E. Lee (eds.), *Chaos, Complexity, and Sociology*, Sage Publications, Thousand Oaks, pp. 30-38.
Milanovic, D. (1997). *Chaos, Criminology, and Social Justice: The New Orderly (Dis)Order*, Praeger Publications, Westport.
Milanovic, D. (2002). *Critical Criminology at the Edge: Postmodern Perspectives, Integration, and Applications*, Praeger Publications, Westport.
Miller, A. (1949). *Death of a Salesman*, Penguin Books, New York.
Minnis, P. E. (1985). *Social Adaptation to Food Stress: A Prehistoric Southwestern Example*, University of Chicago Press, Chicago.
Minsky, M., and Papert, S. (1972). *Artificial Intelligence Progress Report (Ai Memo 252)*, MIT Artificial Intelligence Laboratory, Cambridge.
Mitchell, J., and Young, A. (1775). *American Husbandry. Containing an Account of the Soil, Climate, Production and Agriculture of the British Colonies in North-America and the West-Indies*, J. Bew, London.
Mithen, S. (1996). *The Prehistory of the Mind: The Cognitive Origins of Art, Religion and Science*, Thames and Hudson, London.
Moore, H. L. (1994). *A Passion for Difference: Essays in Anthropology and Gender*, Indiana University Press, Bloomington.
Mortimore, M. (1989). *Adapting to Drought: Farmers, Famines, and Desertification in West Africa*, Cambridge University Press, Cambridge.
Mouzelis, N. (1995). *Sociological Theory: What Went Wrong?* Routledge, London.
Mouzelis, N. (2000), 'The Subjectivist-Objectivist Divide: Against Transcendence', *Sociology*, vol. 34, no. 4, pp. 741-763.
Muller, J. (1997). *Mississippian Political Economy*, Plenum Press, New York.
Munson-Scullin, W., and Scullin, M. (2005), 'Potential Productivity of Midwestern Native American Gardens', *Plains Anthropologist*, vol. 50, no. 193, pp. 9-22.
Murphy, W. P. (1980), 'Secret Knowledge as Property and Power in Kpelle Society', *Africa*, vol. 50, no. 2, pp. 193-207.
Myers, G. (1989). Nineteeth-Century Popularizations of Thermodynamics and the Rhetoric of Social Prophecy, in P. Brantlinger (ed.), *Energy & Entropy: Science and Culture in*

Victorian Britain: Essays from Victorian Studies, Indiana University Press, Bloomington, pp. 307-338.

National Geographic. (2002). *A World Transformed* (September), National Geographic Maps, Washington, DC.

Nedelmann, B. (1990). Georg Simmel as an Analyst of Autonomous Dynamics: The Merry-Go-Round of Fashion, in M. Kaern, B. S. Phillips and R. S. Cohen (eds.), *Georg Simmel and Contemporary Sociology*, Kluwer, Dordrecht, pp. 243-257.

Newman, D. V. (1996), 'Emergence and Strange Attractors', *Philosophy of Science*, vol. 63, no. 2, pp. 245-261.

Nicholas, R. W. (1977). Factions: A Comparative Analysis, in S. W. Schmidt, L. Guasti, C. H. Lande and J. C. Scott (eds.), *Friends, Followers and Factions*, University of California, Berkeley, pp. 55-73.

Nickles, M., Rovatsos, M., and Weiss, G. (eds.). (2004). *Agents and Computational Autonomy. Potential, Risks, Solutions*, Lecture Notes in Artificial Intelligence 2969, a series edited by J. G. Carbonell and J. Siekmann, Springer, Berlin.

Nicolis, G. (1995). *Introduction to Nonlinear Science*, Cambridge University Press, Cambridge.

Nicolis, G., and Prigogine, I. (1977). *Self-Organization in Nonequilibrium Systems: From Dissipative Structures to Order through Fluctuations*, Wiley, New York.

Nicolis, G., and Prigogine, I. (1989). *Exploring Complexity: An Introduction*, W.H. Freeman, New York.

North, M., and Macal, C. (2005). *Managing Business Complexity: Discovering Strategic Solutions with Agent-Based Modelling and Simulation*, Oxford University Press, New York.

Nowak, M. A., May, R. M., and Sigmund, K. (1995), 'The Arithmetics of Mutual Help', *Scientific American*, vol. 272, no. 6, pp. 76-81.

Nye, P. H., and Greenland, D. J. (1960). *The Soil under Shifting Cultivation*, Commonwealth Agricultural Bureaux, Farnham Royal.

O'Brien, M. J., and Holland, T. D. (1992), 'The Role of Adaptation in Archaeological Explanation', *American Antiquity*, vol. 57, no. 1, pp. 36-35.

O'Connor, T. (1994), 'Emergent Properties', *American Philosophical Quarterly*, vol. 31, no. 2, pp. 91-105.

O'Connor, T., and Wong, H. Y. (2002). Emergent Properties, from *The Stanford Encyclopedia of Philosophy (Winter 2002 edition)*, from http://plato.stanford.edu/archives/win2002/entries/properties-emergent/.

O'Neill, E. (1989). *Long Day's Journey into Night*, Yale University Press, New Haven.

Oates, E. D. M. (1968). *Studies in the Ancient History of Northern Iraq*, The British Academy, London.

Oleschko, K., Brambila, R., Brambila, F., Parrot, J-F., and López, P. (2000), 'Fractal Analysis of Teotihuacan, Mexico', *Journal of Archaeological Science*, vol. 27, pp. 1007-1016.

Orbell, J., Morikawa, T., and Allen, N. (2002), 'The Evolution of Political Intelligence: Simulation Results', *British Journal of Political Science*, vol. 32, no. 4, pp. 613-640.

Ortner, S. (1984), 'Theory in Anthropology since the 1960s', *Comparative Studies in Society and History*, vol. 26, pp. 126-166.

Pacheco, O. (2004). Autonomy in an Organizational Context, in M. Nickles, M. Rovatsos and G. Weiss (eds.), *Agents and Computational Autonomy. Potential, Risks, Solutions* (Lecture Notes in Artificial Intelligence 2969), Springer, Berlin, pp. 195-208.

Parker, A. C. (1968). Iroquois Use of Maize and Other Foods, in W. N. Fenton (ed.), *Parker on the Iroquois*, Syracuse University Press, Syracuse, pp. 3-119.

Parsons, E. C. (1928), 'The Laguna Migration to Isleta', *American Anthropologist*, vol. 34, no. 4, pp. 602-613.

Patterson, T. C. (1995). *Toward a Social History of Archaeology in the United States*, Harcourt Brace College Publishers, Fort Worth.

Pauketat, T. (2000). The Tragedy of the Commoners, in M-A. Dobres and J. Robb (eds.), *Agency in Archaeology: Paradigm or Platitude?* Routledge, London, pp. 113-129.

Pendergast, J. F. (1996). *Hamlets Become Villages.* Paper presented at the 23rd Annual Symposium of the Ontario Archaeological Society, Kingston, Ontario.

Pickering, R. B., and Cabrero, M. T. (1998). Mortuary Practices in the Shaft Tomb Region, in R. F. Townsend (ed.), *Ancient West Mexico: Art and Archaeology of the Unknown Past*, Art Institute of Chicago, Chicago, pp. 71-87.

Piha, M. I., and Munns, D. N. (1987), 'Nitrogen Fixation Capacity of Field-Grown Bean Compared to Other Grain Legumes', *Agronomy Journal*, vol. 79, pp. 690-696.

Pines, D. (ed.). (1987). *Emerging Syntheses in Science* (Santa Fe Institute Studies in the Sciences of Complexity, Proceedings Volume I), Addison-Wesley, Reading.

Post, D. G. (2001). Anarchy, State, and the Internet: An Essay on Lawmaking in Cyberspace, in P. Ludlow (ed.), *Crypto Anarchy, Cyberstates, and Pirate Utopias*, MIT Press, Cambridge, pp. 197-211.

Poston, T., and Stewart, I. (1979). *Catastrophe Theory and Its Applications*, Longman Group, London.

Powlson, D. S. (1994). Quantification of Nutrient Cycles Using Long-Term Experiments, in R. A. Leigh and A. E. Johnston (eds.), *Long-Term Experiments in Agricultural and Ecological Sciences*, CAB International, Wallingford, pp. 97-115.

Preucel, R. W., and Hodder, I. (1996). Nature and Culture, in R. W. Preucel and I. Hodder (eds.), *Contemporary Archaeology in Theory: A Reader*, Blackwell, Oxford, pp. 23-38.

Price, B. (1997). The Myth of Post-Modern Science, in R. A. Eve, S. Horsfall and M. E. Lee (eds.), *Chaos, Complexity, and Sociology: Myths, Models, and Theories*, Sage Publications, Thousand Oaks, pp. 3-14.

Prietula, M. J., Carley, K. M., and Gasser, L. (eds.). (1998). *Simulating Organizations: Computational Models of Institutions and Groups*, AAAI Press, Menlo Park.

Prigogine, I. (1961). *Introduction to Thermodynamics of Irreversible Processes* (2d, rev.), Interscience Publishers, New York.

Prigogine, I. (1962). *Non-Equilibrium Statistical Mechanics*, Interscience Publishers, New York.

Prigogine, I. (1971). Unity of Physical Laws and Levels of Description, in M. Grene (ed.), *Interpretations of Life and Mind*, Humanities Press, New York.

Prigogine, I. (1980). *From Being to Becoming: Time and Complexity in the Physical Sciences*, W. H. Freeman, New York.

Prigogine, I. (1998). Pre-Columbian Artifacts and the Post-Newtonian Sciences, in M. Holsbeke and K. Arnaut (eds.), *Offerings for a New Life: Funerary Images from Pre-Columbian West Mexico*, Antwerp Ethnographic Museum, Antwerp, pp. 53-59.

Prigogine, I., Nicolis, G., and Babloyantz, A. (1972a), 'Thermodynamics of Evolution', *Physics Today,* November, pp. 23-28.

Prigogine, I., Nicolis, G., and Babloyantz, A. (1972b), 'Thermodynamics of Evolution: Part II', *Physics Today,* December, pp. 38-44.

Prigogine, I., and Stengers, I. (1984). *Order out of Chaos: Man's New Dialogue with Nature*, Bantam Books, New York.

Purkis, J., and Bowen, J. (2004). Introduction: Why Anarchism Still Matters, in J. Purkis and J. Bowen (eds.), *Changing Anarchism: Anarchist Theory and Practice in a Global Age*, Manchester University Press, Manchester, pp. 1-19.

Ramos de la Vega, J., and López Mestas C. L. (1996), 'Datos Preliminares Sobre El Descubrimiento De Una Tumba De Tiro Un El Sitio De Huitzilapa, Jalisco', *Ancient Mesoamerica*, vol. 7, no. 1, pp. 121-134.

Rasmussen, P. E., Goulding, K. W. T., Brown, J. R., Grace, P. R., Janzen, H. H., and Körschens, M. (1998), 'Long-Term Agroecosystem Experiments: Assessing Agricultural Sustainability and Global Change', *Science*, vol. 282, no. 30, pp. 893-896.

Rautman, A. (1998), 'Hierarchy and Heterarchy in the American Southwest: A Comment on Mcguire and Saitta', *American Antiquity*, vol. 63, no. 2, pp. 325-333.

Redman, C. L. (1999). *Human Impact on Ancient Environments*, University of Arizona Press, Tucson.

Reed, M., and Harvey, D. L. (1992), 'The New Science and the Old: Complexity and Realism in the Social Sciences', *Journal for the Theory of Social Behaviour*, vol. 22, pp. 353-380.

Renfrew, C. (1978), 'Trajectory Discontinuity and Morphogenesis: The Implications of Catastrophe Theory for Archaeology', *American Antiquity*, vol. 43, no. 2, pp. 203-222.

Renfrew, C. (1979). Systems Collapse as Social Transformation - Catastrophe and Anastrophe in Early State Societies, in C. Renfrew and K. L. Cooke (eds.), *Transformations: Mathematical Approaches to Culture Change*, Academic Press, New York, pp. 481-506.

Renfrew, C., and Cooke, K. L. (eds.). (1979). *Transformations: Mathematical Approaches to Culture Change*, Academic Press, New York.

Reynolds, C. W. (1987), 'Flocks, Herds, and Schools', *Computer Graphics*, vol. 21, no. 4, pp. 25-34.

Reynolds, R., and Kobti, Z. (2003, April 9th-13th, 2003). *The Effect of Environmental Variability on the Resilience of Social Networks: An Example Using the Mesa Verde Pueblo Culture.* Paper presented at the Society for American Archaeology Annual Meetings, Montreal.

Richards, D. (1996). From Individuals to Groups: The Aggregation of Votes and Chaotic Dynamics, in L. D. Kiel and E. Elliott (eds.), *Chaos Theory in the Social Sciences. Foundations and Applications*, University of Michigan Press, Ann Arbor, pp. 89-116.

Richards, D. (ed.). (2000). *Political Complexity. Nonlinear Models of Politics*, University of Michigan Press, Ann Arbor.

Ridley, M. (1996). *Evolution* (2nd edition), Blackwell Science, Cambridge.

Robbins, J. (2001), 'Ritual Communication and Linguistic Ideology: A Reading and Partial Reformulation of Rappaport's Theory of Ritual', *Current Anthropology*, vol. 42, pp. 591-614.

Rogers, R. J. (1995). Tribes as Heterarchy: A Case Study from the Prehistoric Southeastern United States, in M. E. Robert, C. L. Crumley and J. E. Levy (eds.), *Heterarchy and the Analysis of Complex Societies* (Archaeological Papers of the American Anthropological Association No. 6), American Anthropological Association, Washington, DC, pp. 7-16.

Ronfeldt, D. (1996). *Tribes, Institutions, Markets, Networks: A Framework About Societal Evolution*, RAND, Santa Monica.

Root, D. (1983). Information Exchange and the Spatial Configurations of Egalitarian Societies, in J. A. Moore and A. S. Keene (eds.), *Archaeological Hammers and Theories*, Academic Press, New York, pp. 193-219.

Rost, J. C. (1991). *Leadership for the Twenty-First Century*, Praeger, Westport.

Rothenbuhler, E. W. (1998). *Ritual Communication: From Everyday Conversation to Mediated Ceremony*, Sage Publications, Thousand Oaks.

Roux, V. (2003), 'A Dynamic Systems Framework for Studying Technological Change: Application to the Emergence of the Potter's Wheel in the Southern Levant', *Journal of Archaeological Method and Theory*, vol. 10, no. 1, pp. 1-30.

Rowling, J. K. (1997). *Harry Potter and the Sorcerer's Stone*, Scholastic Press, New York.

Rudin, N. (1997). *Dictionary of Modern Biology*, Barron's Educational Series, Inc., Haupapauge.

Rueger, A. (2000), 'Robust Supervenience and Emergence', *Philosophy of Science*, vol. 67, no. 3, pp. 466-489.

Ruggles, A. J., and Church, R. L. (1996). Spatial Allocation in Archaeology: An Opportunity for Reevaluation, in H. D. G. Maschner (ed.), *New Methods Old Problems: Geographic Information Systems in Modern Archaeological Research* (Occasional Paper No. 23), Carbondale Center for Archaeological Investigations, Carbondale, pp. 150-161.

Rushforth, S., and Upham, S. (1992). *A Hopi Social History*, University of Texas Press, Austin.

Russell, E. J., and Russell, E. W. (1973). *Soil Conditions and Plant Growth* (10th), Longman, London.

Rutman, D. B. (1967). *Husbandmen of Plymouth: Farms and Villages in the Old Colony, 1620-1692*, Beacon Press, Boston.

Sagard, G. (1939). *The Long Journey to the Country of the Hurons* (H. H. Langton, Trans.), The Champlain Society, Toronto.

Sahlins, M. D., and Service, E. E. (eds.). (1960). *Evolution and Culture*, University of Michigan Press, Ann Arbor.

Saitta, D. (1991), 'Room Use and Community Organization at the Pettit Site, West Central New Mexico', *Kiva*, vol. 56, pp. 383-409.

Saitta, D. (1994). The Political Economy and Ideology of Early Population Aggregation in the Togeye Canyon, A.D. 1150-1250, in T. L. Howell and T. Stone (eds.), *Exploring Social, Political and Economic Organization in the Zuni Region, Anthropological Research Papers No. 46*, Arizona State University, Tempe, pp. 47-60.

Salisbury, R. F., and Silverman, M. (1977). Introduction: Factions and the Dialectic, in M. Silverman and R. F. Salisbury (eds.), *A House Divided? Anthropological Studies of Factionalism* (Social and Economic Papers No. 9, Institute of Social and Economic Research, Memorial University of Newfoundland), University of Toronto, Toronto, pp. 1-20.

Sallaberger, W., and Ur, J. A. (2004). Tell Beydar / Nabada in Its Regional Setting, in L. Milano, W. Sallaberger, P. Talon and K. V. Lerberghe (eds.), *Third Millennium Cuneiform Texts from Tell Beydar (Seasons 1996-2002)*, Brepols, Turnhout, pp. 55-71.

Samford, P. (2000). *Power Flows in Many Channels: Pits and West African-Based Spiritual Traditions in Colonial Virginia.* Unpublished Ph.D. dissertation, Department of Anthropology, University of North Carolina.

Sandor, J. A. (1995). Searching Soil for Clues About Southwest Prehistoric Agriculture, in H. W. Toll (ed.), *Soil, Water, Biology, and Belief in Prehistoric and Traditional Southwestern Agriculture*, New Mexico Archaeological Council, Albuquerque, pp. 119-137.

Sandor, J. A., and Eash, N. S. (1991), 'Significance of Ancient Agricultural Soils for Long-Term Agronomic Studies and Sustainable Agriculture Research', *Agronomy Journal,* vol. 83, no. 1, pp. 29-37.

Sandor, J. A., and Gersper, P. L. (1988), 'Evaluation of Soil Fertility in Some Prehistoric Agricultural Terraces in New Mexico', *Agronomy Journal*, vol. 80, no. 5, pp. 846-850.

Saperstein, A. M. (1997). The Origins of Order and Disorder in Physical and Social Deterministic Systems, in R. A. Eve, S. Horsfall and M. E. Lee (eds.), *Chaos, Complexity, and Sociology*, Sage Publications, Thousand Oaks, pp. 102-124.

Sawyer, R. K. (2001), 'Emergence in Sociology: Contemporary Philosophy of Mind and Some Implications for Sociological Theory', *American Journal of Sociology*, vol. 107, no. 3, pp. 551-585.

Sawyer, R. K. (2004), 'The Mechanisms of Emergence', *Philosophy of the Social Sciences*, vol. 34, no. 2, pp. 260-282.

Scarborough, V. L., Valdez, F., and Dunning, N. (eds.). (2003). *Heterarchy, Political Economy, and the Ancient Maya: The Three Rivers Region of the East-Central Yucatan Peninsula*, University of Arizona Press, Tucson.

Schieve, W. C., and Allen, P. M. (eds.). (1982). *Self-Organization and Dissipative Structures: Applications in the Physical and Social Sciences*, University of Texas Press, Austin.

Schillo, M., and Fischer, K. (2004). A Taxonomy of Autonomy in Multiagent Organisation, in M. Nickles, M. Rovatsos and G. Weiss (eds.), *Agents and Computational Autonomy. Potential, Risks, Solutions* (Lecture Notes in Artificial Intelligence 2969), Springer, Berlin, pp. 68-82.

Schillo, M., and Fischer, K. (n.d.). Holonic Multiagent Systems, in *Zeitschrift Für Künstliche Intelligenz, No. 3*, From http://www.millo.de/virtosphere/data/publications/articles/SchilloFischer.HoloMAS.pdf.

Schillo, M., Fischer, K., and Klein, C. T. (2001). The Micro-Macro Link in DAI and Sociology, in S. Moss and P. Davidsson (eds.), *Multiagent-Based Simulation: Second International Workshop on Multiagent-Based Simulation Boston MA, USA, July, 2000* (Lecture Notes in Artificial Intelligence 2969), Springer, Berlin.

Schillo, M., Fley, B., Florian, M., Hillebrandt, F., and Hinck, D. (2002). Self-Organization in Multiagent Systems, in *Proceedings of the Third International Workshop on Modelling Artificial Societies and Hybrid Organizations (MASHO).* from http://www.millo.de/virtosphere/data/publications/workshops/2002/Schillo+._MASHO_2002.pdf.

Schloen, D. (2001). *The House of the Father as Fact and Symbol: Patrimonialism in Ugarit and the Ancient Near East*, Eisenbrauns, Winona Lake.

Schmidt, S. W., Guasti, L., Lande, C. H., and Scott, J. C. (eds.). (1977). *Friends, Followers and Factions*, University of California, Berkeley.

Schoenfelder, J. W. (2003). *Negotiating Poise in a Multi-Hierarchical World: An Archaeological Exploration of Irrigated Rice Agriculture, Ideology, and Political Balances in the Coevolution of Intersecting Complex Networks in Bali.* Unpublished Ph.D. dissertation, Department of Anthropology, The University of California, Los Angeles.

Schofield, N. (1995). Chaos or Equilibrium in a Political Economy, in A. Albert (ed.), *Chaos and Society*, IOS Press, Amsterdam, pp. 193-209.

Schroder, J. (1998), 'Emergence: Non-Deducibility or Downwards Causation?' *The Philosophical Quarterly*, vol. 48, no. 193, pp. 433-452.

Schrödinger, E. (1944). *What Is Life?* Cambridge University Press, New York.

Schulz, C. M. (2004). *The Complete Peanuts: 1950 to 1952*, Fantographics Books, Seattle.

Schweizer, T. (1997), 'Embeddedness of Ethnographic Cases: A Social Networks Perspective', *Current Anthropology*, vol. 38, pp. 739-760.

Scott, G. P. (ed.). (1991). *Time, Rhythms, and Chaos in the New Dialogue with Nature*, University of Iowa Press, Ames.

Scott, J. (2001). Where Is Social Structure?, in J. López and G. Potter (eds.), *After Postmodernism: An Introduction to Critical Realism*, Athlone Press, London, pp. 77-85.

Scott, J. C. (1987). *Weapons of the Weak: Everyday Forms of Peasant Resistance*, Yale University Press, New Haven.

Scott, J. C. (1998). *Seeing Like a State: How Certain Schemes to Improve the Human Condition Have Failed*, Yale University Press, New Haven.

Service, E. R. (1962). *Primitive Social Organization: An Evolutionary Perspective*, Random House, New York.

Sewell, Jr., W. H. (1992), 'A Theory of Structure: Duality, Agency, and Transformation', *American Journal of Sociology*, vol. 98, no. 1, pp. 1-29.

Shakespeare, W. (1992). *Hamlet, Oxford School Shakespeare*, Oxford University Press, Oxford.

Shanks, M., and Tilley, C. (1987). *Social Theory and Archaeology*, University of New Mexico Press, Albuquerque.

Shawcross, W. (1972). Energy and Ecology: Thermodynamic Models in Archaeology, in D. L. Clarke (ed.), *Models in Archaeology*, Methuen, London, pp. 577-622.

Shermer, M. (1995), 'Exorcising Laplace's Demon: Chaos and Antichaos, History and Metahistory', *History and Theory*, vol. 34, no. 1, pp. 59-83.

Shoemaker, S. (2002), 'Kim on Emergence', *Philosophical Studies,* vol. 108, pp. 53-63.

Silliman, S. (2001), 'Agency, Practical Politics and the Archaeology of Culture Contact', *Journal of Social Archaeology*, vol. 1, no. 2, pp. 190-209.

Silverman, H. (2002). *Ancient Nasca Settlement and Society*, University of Iowa Press, Iowa City.

Silverman, M., and Salisbury, R. F. (eds.). (1977). *A House Divided?* Anthropological Studies of Factionalism. Social and Economic Papers No. 9, Institute of Social and Economic Research, Memorial University of Newfoundland, University of Toronto, Toronto.

Simmel, G. (1971 [1908]). Group Expansion and the Development of Individuality, in D. Levine (ed.), *Georg Simmel: On Individuality and Social Forms*, University of Chicago Press, Chicago, pp. 251-293.

Smith, A. (2001), 'The Limitations of Doxa. Agency and Subjectivity from an Archaeological Point of View', *Journal of Social Archaeology*, vol. 1, no. 2, pp. 155-171.

Smith, A. (2003). *The Political Landscape. Constellations of Authority in Ancient Complex Polities*, University of California Press, Berkeley.

Smith, E. A., and Winterhalder, B. (1992). *Evolutionary Ecology and Human Behavior*, Aldine de Gruyter, New York.

Smith, T. S. (1997). Nonlinear Dynamics and the Micro-Macro Bridge, in R. A. Eve, S. Horsfall and M. E. Lee (eds.), *Chaos, Complexity, and Sociology*, Sage Publications, Thousand Oaks, pp. 52-63.

Smith, T. S., and Stevens, G. T. (1996), 'Emergence, Self-Organization, and Social Interaction: Arousal-Dependent Structure in Social Systems', *Sociological Theory*, vol. 14, pp. 131-153.

Soltis, J., Boyd, R., and Richerson, P. J. (1995), 'Can Group-Functional Behaviors Evolve by Cultural Group Selection? An Empirical Test', *Current Anthropology*, vol. 36, no. 3, pp. 473-494.

Somers, M. R. (1998), '"We're No Angels": Realism, Rational Choice, and Relationality in Social Science', *The American Journal of Sociology*, vol. 104, no. 3, pp. 722-784.

Spencer-Wood, S. M. (1990). *Non-Linear Dynamics in Cultural Processes: Implications for Theories of Culture and Cultural Evolution.* Paper presented at the Conference on Dynamical Description and Human Systems, University of Cambridge.

Spencer-Wood, S. M. (1996). Cultural Complexity, Non-Linear Systems Theory, and Multi-Scalar Analysis, in D. A. Meyer, P. C. Dawson and D. T. Hanna (eds.), *Debating Complexity, Proceedings of the 26th Annual Chacmool Conference*, Archaeological Association of the University of Calgary, Calgary, pp. 54-63.

Spencer-Wood, S. M. (2000). Strange Attractors: Feminist Theory, Non-Linear Systems Theory, and Their Implications for Archaeological Theory, in M. B. Schiffer (ed.), *Social Theory in Archaeology*, University of Utah Press, Salt Lake City, pp. 112-125.

Sperry, R. W. (1986), 'Macro- Versus Micro-Determinism', *Philosophy of Science*, vol. 53, pp. 265-270.

Sperry, R. W. (1991), 'In Defense of Mentalism and Emergent Interaction', *Journal of Mind and Behavior*, vol. 12, pp. 221-245.

Stahle, D. W., and Cleveland, M. K. (1994), 'Tree-Ring Reconstructed Rainfall over the Southeastern USA during the Medieval Warm Period and Little Ice Age', *Climatic Change*, vol. 26, pp. 199-212.

Stark, D. (2001). Ambiguous Assets for Uncertain Environments: Heterarchy in Postsocialist Firms, in P. DiMaggio (ed.), *The Twenty-First-Century Firm: Changing Economic Organization in International Perspective*, Princeton University Press, Princeton, pp. 69-104.

Steffen, W. L., Sanderson, A., Tyson, P. D., Jager, J., Matson, P. A., Moore III, B., et al. (2004). *Global Change and the Earth System: A Planet under Pressure*, Springer-Verlag, Heidelberg.

Stein, G. (1994). Segmentary States and Organizational Variation in Early Complex Societies: A Rural Perspective, in G. M. Schwartz and S. A. Falconer (eds.), *Archaeological Views from the Countryside*, Smithsonian Institution Press, Washington, DC, pp. 10-18.

Stein, G. (1998), 'Heterogeneity, Power, and Political Economy: Some Current Research Issues in the Archaeology of Old World Complex Societies', *Journal of Archaeological Research*, vol. 6, no. 1, pp. 1-44.

Stein, G., and Wattenmaker, P. (1990). The 1987 Tell Leilan Regional Survey: Preliminary Report, in N. Miller (ed.), *Economy and Settlement in the Near East: Analyses of Ancient Sites and Materials* (MASCA Research Papers in Science and Archaeology No. 7), University Museum, Philadelphia, pp. 8-18.

Steward, J. (1960), 'Review of White's *The Evolution of Culture*', *American Anthropologist,* vol. 62, pp. 144-148.

Stewart, B., and Lockyer, J. N. (1868), 'The Place of Life in a Universe of Energy', *Macmillan's*, vol. 20, p. 319.

Stone, G. D., and Downum, C. E. (1999), 'Non-Boserupian Ecology and Agricultural Risk: Ethnic Politics and Land Control in the Arid Southwest', *American Anthropologist*, vol. 101, pp. 113-128.

Stone, T. (1994). The Process of Aggregation in the Zuni Region: Reasons and Implications, in T. L. Howell and T. Stone (eds.), *Exploring Social, Political and Economic Organization in the Zuni Region* (Anthropological Research Paper No. 47), Arizona State University, Tempe, pp. 9-24.

Stone, T. (1999), 'The Chaos of Collapse: Disintegration and Reintegration of Inter-Regional Systems', *Antiquity*, vol. 73, pp. 110-118.

Stone, T. (2002), 'Kiva Diversity in the Point of Pines Region of Arizona', *Kiva*, vol. 67, pp. 385-412.

Stone, T. (2003), 'Social Identity and Ethnic Interaction in the Western Pueblos of the American Southwest', *Journal of Archaeological Method and Theory*, vol. 10, pp. 31-67.

Stone, T., and Howell, T. L. (1993). Contemporary Theory in the Study of Socio-Political Organization, in T. L. Howell and T. Stone (eds.), *Exploring Social, Political, and Economic Organization in the Zuni Region* (Anthropological Research Paper No. 47), Arizona State University, Tempe, pp. 103-110.

Stones, R. (2001), 'Refusing the Realism-Structuration Divide', *European Journal of Social Theory*, vol. 4, no. 2, pp. 177-197.

Stuart, G. (2003). *Pre-Hispanic Sociopolitical Development and Wetland Agriculture in the Tequila Valleys of West Mexico.* Unpublished Ph.D. dissertation, Department of Anthropology, Arizona State University, Tempe.

Suleiman, R., Troitzsch, K. G., and Gilbert, N. (eds.). (2000). *Tools and Techniques for Social Science Simulation*, Physica-Verlag, Heidelberg.

Sumner, W. M. (1989), 'Population and Settlement Area: An Example from Iran', *American Anthropologist*, vol. 91, pp. 631-641.

Sztompka, P. (1991). *Society in Action: The Theory of Social Becoming*, Polity Press, Cambridge.

Sztompka, P. (2000). *Trust: A Sociological Theory*, Cambridge University Press, Cambridge.

Sztompka, P. (ed.). (1994). *Agency and Structure. Reorienting Social Theory*, Gordon and Breach, Switzerland.

Tainter, J. A. (1988). *The Collapse of Complex Societies*, Cambridge University Press, Cambridge.

Thom, R. (1975). *Structural Stability and Morphogenesis: An Outline of a General Theory of Models*, M.A. Benjamin, Reading.

Thwaites, R. J. (ed.). (1896-1901). *The Jesuit Relations and Allied Documents*, Burrows Brothers, Cleveland.

Titiev, M. (1944). Old Oraibi, a Study of the Hopi Indians of Third Mesa, in *Papers of the Peabody Museum of American Archaeology and Ethnology, Vol. 22, No. 1*, Harvard University, Cambridge.

Tooker, E. (1964). An Ethnography of the Huron Indians, 1615-1649, in *Bureau of American Ethnology Bulletin 190*, Smithsonian Institution, Washington, DC.

Townsend, R. F. (1998). Before Gods, Before Kings, in R. F. Townsend (ed.), *Ancient West Mexico: Art and Archaeology of the Unknown Past*, Art Institute of Chicago, Chicago, pp. 107-135.

Turner, F. (1997). Foreword: Chaos and Social Science, in R. A. Eve, S. Horsfall and M. E. Lee (eds.), *Chaos, Complexity, and Sociology: Myths, Models, and Theories*, Sage Publications, Thousand Oaks, pp. xi-xxvii.

Twain, M. (1999). *Adventures of Huckleberry Finn* (3rd edition), W.W. Norton and Company, New York.

Tyndall, G. (n.d.). A Preliminary Study of Feasting at Navajas Using the Circle 5 Ceramic Assemblage, in C. S. Beekman and P. C. Weigand (eds.), *La Tradición Teuchitlan: Un Primer Intento a Síntesis*, Colegio de Michoacan, Zamora.

Upham, S., MacNeish, R. S., Galinat, W. C., and Stevenson, C. M. (1987), 'Evidence Concerning the Origin of Maiz De Ocho', *American Anthropologist*, vol. 89, no. 2, pp. 410-419.

Upham, S., MacNeish, R. S., and Stevenson, C. M. (1988), 'The Age and Evolutionary Significance of Southwestern Maiz De Ocho', *American Anthropologist*, vol. 90, pp. 683-684.

Ur, J. A. (2003), 'Corona Satellite Photography and Ancient Road Networks: A Northern Mesopotamian Case Study', *Antiquity*, vol. 77, pp. 102-115.

Ur, J. A. (2004). *Urbanism and Society in the Third Millennium Upper Khabur Basin.* Unpublished Ph. D. dissertation, Near Eastern Languages and Civilizations, University of Chicago.

Uribelarrea, M., Below, F., and Moose, S. (2004), 'Grain Composition and Productivity of Maize Hybrids Derived from the Illinois Protein Strains in Response to Variable Nitrogen Supply', *Crop Science*, vol. 44, no. 5, pp. 1593-1600.

Valeri, V. (2001). *Fragments from Forests and Libraries*, Carolina Academic Press, Durham, N.C.

Van der Leeuw, S. E. (ed.). (1981a). *Archaeological Approaches to the Study of Complexity*, Universiteit van Amsterdam, Amsterdam.

Van der Leeuw, S. E. (1981b). Information Flows, Flow Structures and the Explanation of Change in Human Institutions, in S. E. van der Leeuw (ed.), *Archaeological Approaches to the Study of Complexity*, Universiteit van Amsterdam, Amtersdam, pp. 230-329.

Van der Leeuw, S. E., and McGlade, J. (eds.). (1997a). *Time, Process, and Structured Transformation in Archaeology*, Unwin Hyman Press, London.

Van der Leeuw, S. E., and McGlade, J. (1997b). Structural Change and Bifurcation in Urban Evolution: A Non-Linear Dynamical Perspective, in S. E. van der Leeuw and J. McGlade (eds.), *Time, Process and Structured Transformation in Archaeology*, Routledge, London, pp. 331-372.

Van der Leeuw, S. E., and Torrence, R. (eds.). (1989). *What's New? A Closer Look at the Process of Innovation*, Unwin Hyman, London.

Vaughn, D. (1999), 'The Role of the Organization in the Production of Techno-Scientific Knowledge', *Social Studies of Science*, vol. 29, no. 6, pp. 913-943.

Von Bertalanffy, L. (1951). *General System Theory: A New Approach to Unity of Science* (Revised), John Hopkins Press, New York.

Von Winning, H. (1969). *Pre-Columbian Art of Mexico and Central America*, Thames and Hudson, London.

Waldrop, M. (1992). *Complexity: The Emerging Science at the Edge of Chaos*, Simon and Schuster, New York.

Wallace, A. F. C. (1970). *Culture and Personality* (2nd edition), Random House, New York.

Wallerstein, I. (1997), 'Social Science and the Quest for a Just Society', *American Journal of Sociology*, vol. 102, pp. 1241-1257.

Watson, P. J., LeBlanc, S. A., and Redman, C. (1971). *Explanation in Archaeology: An Explicitly Scientific Approach*, Columbia University Press, New York.

Watts, D. J. (1999). *Small Worlds: The Dynamics of Networks between Order and Randomness*, Princeton University Press, Princeton.

Weigand, P. C. (1985). Evidence for Complex Societies During the Western Mesoamerican Classic Period, in M. S. Foster and P. C. Weigand (eds.), *The Archaeology of West and Northwest Mesoamerica*, Westview Press, Boulder, pp. 47-91.

Weir, W. W. (1926). A Study of the Value of Crop Rotation in Relation to Soil Productivity. United States Department of Agriculture, Technical Bulletin No. 1377.

Wellman, B., and Frank, K. A. (2001). Network Capital in a Multilevel World: Getting Support from Personal Communities, in N. Lin, K. Cook and R. S. Burt (eds.), *Social Capital: Theory and Research*, Aldine de Gruyter, New York, pp. 233-272.

Welshon, R. (2002), 'Emergence, Supervenience, and Realization', *Philosophical Studies*, vol. 108, pp. 39-51.

Wheatley, M. J. (1994). *Leadership and the New Science: Learning About Organization from an Orderly Universe*, Berrett-Kohler Publishers Inc., San Francisco.

White, L. A. (1943), 'Energy and the Evolution of Culture', *American Anthropologist*, vol. 43, pp. 335-356.

White, L. A. (1949). *The Science of Culture: A Study of Man and Civilization*, Farrar Straus, New York.

White, L. A. (1954). The Energy Theory of Cultural Development, in K. M. Kapadia (ed.), *The Ghurye Felicitation Volume*, Popular Book Depot, Bombay, pp. 1-10.

White, L. A. (1959). *The Evolution of Culture: The Development of Civilization to the Fall of Rome*, McGraw-Hill, New York.

Whiteley, P. M. (1988). *Deliberate Acts: Changing Hopi Culture through the Oraibi Split*, University of Arizona Press, Tucson.

Whiten, A. J. G., McGrew, W. C., Nishida, T., Reynolds, V., Sugiyama, Y., Tutin, C. E. G., et al. (1999), 'Culture in Chimpanzees', *Nature*, vol. 399, pp. 682-685.

Whitmore, T. M., Turner II, B. L., Johnson, D. L., Kates, R. W., and Gottschang, T. R. (1990). Long-Term Population Change, in B. L. Turner II, W. C. Clark, R. W. Kates, J. F. Richards, J. T. Mathews and W. B. Meyer (eds.), *The Earth as Transformed by Human Action*, Cambridge University Press, Cambridge, pp. 25-39.

Wiener, N. (1948). *Cybernetics; or, Control and Communication in the Animal and the Machine*, J. Wiley, New York.

Wilkinson, T. J. (1982), 'The Definition of Ancient Manured Zones by Means of Extensive Sherd-Sampling Techniques', *Journal of Field Archaeology*, vol. 9, pp. 323-333.

Wilkinson, T. J. (1993), 'Linear Hollows in the Jazira, Upper Mesopotamia', *Antiquity*, vol. 67, pp. 548-562.

Wilkinson, T. J. (1994), 'The Structure and Dynamics of Dry-Farming States in Upper Mesopotamia', *Current Anthropology*, vol. 35, pp. 483-520.

Wilkinson, T. J. (2003). *Archaeological Landscapes of the Near East*, University of Arizona Press, Tucson.

Wilkinson, T. J., Christiansen, J. H., Widell, M., Ur, J., and Altaweel, M. (n.d.), 'Urbanization within a Dynamic Environment: Modelling Bronze Age Communities in Upper Mesopotamia', *American Anthropologist*, (in press).

Will, G. F., and Hyde, G. E. (1917). *Corn among the Indians of the Upper Missouri*, The William Harvey Miner Co., St. Louis.

Willis, K. J., Gillson, L., and Brncic, T. M. (2004), 'How "Virgin" Is Virgin Rainforest?', *Science*, vol. 304, no. 5669, pp. 402-440.

Willmott, R. (1997), 'Structure, Culture and Agency: Rejecting the Current Orthodoxy of Organisation Theory', *Journal for the Theory of Social Behaviour*, vol. 27, pp. 93-123.

Willmott, R. (1999), 'Structure, Agency and the Sociology of Education: Rescuing Analytical Dualism', *British Journal of Sociology of Education*, vol. 20, pp. 5-21.

Wilson, D. B. (1989). A Physicist's Alternative to Materialism: The Religious Thought of George Gabriel Stokes, in P. Brantlinger (ed.), *Energy & Entropy: Science and Culture in Victorian Britain: Essays from Victorian Studies*, Indiana University Press, Bloomington, pp. 178-204.

Wilson, E. O. (1998). *Consilience: The Unity of Knowledge*, Random House, New York.

Wimsatt, W. C. (1976). Reductionism, Levels of Organization, and the Mind-Body Problem, in G. C. Globus, G. Maxwell and I. Savodnik (eds.), *Consciousness and the Brain: A Scientific and Philosophical Inquiry*, Plenum, New York, pp. 205-267.

Wimsatt, W. C. (1997). Aggregativity: Reductive Heuristics for Finding Emergence, in *Proceedings of the 1996 Biennial Meetings of the Philosophy of Science Association. Part II: Symposia Papers* (Philosophy of Science 64, supplement: S372-384).

Yaeger, J. (2000). Introducing an Archaeology to Communities, in M. A. Canuto and J. Yaeger (eds.), *The Archaeology of Communities: A New World Perspective*, Routledge, London, pp. 1-15.

Yerkes, R. W. (2005), 'Bone Chemistry, Body Parts, and Growth Marks: Evaluating Ohio Hopewell and Cahokia Mississippian Seasonality, Subsistence, Ritual, and Feasting', *American Antiquity*, vol. 70, no. 2, pp. 241-264.

Yoffee, N. (2005). *Myths of the Archaic State: Evolution of the Earliest Cities, States, and Civilizations*, Cambridge University Press, Cambridge.

Yukl, G. (1999), 'An Evaluation of Conceptual Weaknesses in Transformational and Charismatic Leadership Theories', *Leadership Quarterly*, vol. 10, no. 2, pp. 285-305.

Zimansky, P. (2005). Archaeology and Texts in the Ancient Near East, in S. Pollock and R. Bernbeck (eds.), *Archaeologies of the Middle East: Critical Perspectives*, Blackwell Publishing, Oxford, pp. 308-326.

Zubrow, E. (1985), 'Fractals, Cultural Behaviour, and Prehistory', *American Archeology*, vol. 5, no. 1, pp. 63-77.

Index